ROUTLEDGE LIBRARY EDITIONS: URBANIZATION

Volume 1

URBAN AMERICA EXAMINED

URBAN AMERICA EXAMINED

A Bibliography

DALE E. CASPER

Routledge
Taylor & Francis Group

LONDON AND NEW YORK

First published in 1985 by Garland Publishing Inc.

This edition first published in 2018
by Routledge
2 Park Square, Milton Park, Abingdon, Oxon OX14 4RN

and by Routledge
605 Third Avenue, New York, NY 10017

Routledge is an imprint of the Taylor & Francis Group, an informa business

© 1985 Dale E. Casper

British Library Cataloguing in Publication Data
A catalogue record for this book is available from the British Library

ISBN: 978-0-8153-8014-6 (Set)
ISBN: 978-1-351-21390-5 (Set) (ebk)
ISBN: 978-0-8153-7915-7 (Volume 1) (hbk)
ISBN: 978-1-351-21666-1 (Volume 1) (ebk)

Publisher's Note
The publisher has gone to great lengths to ensure the quality of this reprint but points out that some imperfections in the original copies may be apparent.

Disclaimer
The publisher has made every effort to trace copyright holders and would welcome correspondence from those they have been unable to trace.

URBAN AMERICA EXAMINED
A Bibliography

Dale E. Casper

GARLAND PUBLISHING INC. · NEW YORK & LONDON
1985

Library of Congress Cataloging in Publication Data

Casper, Dale E.
  Urban America examined.

  (Garland reference library of social science ; v. 269)
  Includes index.
  1. Cities and town—United States—Bibliography.
2. Urbanization—United States—Bibliography.
3. Sociology, Urban—Bibliography. I. Title.
II. Series.
Z7164.U7C375 1985   016.3077′6′0973   84-48384
[HT123]
ISBN 0-8240-8815-8 (alk. paper)

Printed on acid-free, 250-year-life paper
Manufactured in the United States of America

To Helen
For Her Love and Understanding

CONTENTS

Preface   ix

Introduction to the Literature of Urban Studies:
Highlights, Trends, and Directions   3

Guide to Use of This Book   13

Abstracts and Indexes   17

Periodicals   21

Geographic Quintet   29

    United States (general)   31

    The East   47

    The South   61

    The Midwest   73

    The West   87

An Ambience of Ethnicity   99

"Sewer" Socialism   129

All God's People in the City   139

Politics as Usual   155

Schooling the City   177

Moving Goods and People   191

Geographic Index   203

Topical Index   211

PREFACE

Important contributions have been made in the near and distant past to the organization and presentation of the bibliographic history of urban studies. I have consulted many of these works and wish to express my admiration of these efforts and acknowledge their value to the creation of this bibliography. I wish also to note my appreciation for the efforts made by past urban librarians at Michigan State University whose diligence in collection development provided me with access to many resources useful to this bibliographic project which are not available in other academic libraries. Finally, I extend sincere thanks to Henry Koch for his understanding of the time commitment necessary for the successful completion of this bibliography.

URBAN AMERICA EXAMINED

INTRODUCTION TO THE
LITERATURE OF URBAN STUDIES:
HIGHLIGHTS, TRENDS, AND DIRECTIONS

Theory

Though their theories have been greatly modified since
they were first introduced, Karl Marx, Max Weber, and Emile
Durkheim continue to be the social theorists whose works influence and guide most researchers of social phenomena, including research into the urban environment. An examination
of these three distinct schools of social theory will be
useful in understanding the literature of urban studies.

To begin, it must be understood that the primary focus of
the three theorists (Marx, Weber, and Durkheim) was the social
and political implications of the development of capitalism.
Each of the three did not develop a theory of the city but
rather each developed an explanation of the changing basis of
social relations brought about by the development of capitalism. The difference between these three theorists is seen
in the particular change in social relations identified in
their theories. Karl Marx believed that cities provided an
illustration of the forces unleashed by the development of
capitalist production. Max Weber, however, identified the
change in social relations in cities with the promotion of the
growth of calculative rationality. Emile Durkheim, on the
other hand, maintained that cities promoted the disintegration
of moral cohesion.

All three writers believed that the city was significant
only in the context of a transition from feudal society to
capitalism. Max Weber insisted that cities broke the political
and economic relations of feudalism and thereby established a
new spirit of rationality which was necessary to capitalist
entrepreneurship and democratic rights for citizens. Emile
Durkheim agreed that cities broke the bonds of feudalism and
its traditional morality but saw the outcome of this action
to be the growth of the division of labor in society. For
Karl Marx, the division between medieval cities and their
immediate surrounding countryside was an expression of the
antithesis between developing capitalism and the older feudal

society. Each of the theorists agreed that the city was important in understanding social relations only during the
medieval period. The modern city, however, cannot be analyzed
in the same terms because it is no longer the basis of human
association according to Weber, the locus of the division of
labor according to Durkheim, nor the expression of a specific
mode of production according to Marx.

But the city for all three theorists does retain its influence on the development of fundamental social processes
created within a capitalist society. The city is not a cause
but a significant condition for certain developments in social
relations. Marx argued that the city was an important condition
of the self-realization of the proletariat as a politically
and economically organized class in opposition to the bourgeoisie,
even if the city did not create the proletariat in Marx's view.
Durkheim found urbanization to be an important precondition for
the development of the division of labor. For Max Weber, the
medieval city was an association of individuals which fostered
the development of economic rationality and encouraged a direct
challenge to feudal political power which subsequently formed
the basis of modern democratic forms of government.

Thus, each of the major creators of social theory whose
works are studied by many researchers of social phenomena
concluded that the contemporary city was not an object fruitful
for theoretical analysis. Theory is, therefore, not a leading
component found in the structure of urban research. Research
in urbanization, urbanism, and urban development is devoid of
firm theoretical foundations. Research into the city and its
various characteristics has been more analytic in nature and
when research has been analytic, it has been more case-study
oriented than general-rule conducive. Comparative studies are
the closest that research into the urban environment has come
to providing generalizations which could serve as theory if
retained and tested. Theory as guide is not a common thread
in urban-related research.

Research and Methodology

Urban development as an insignificant characteristic of
the national history was an unacceptable premise for early
American urban researchers. Urban growth in America had been
seen in late-nineteenth-century research as an integral part
of the American nation which was defined by its democratic
ideology and republican institutions. In contrast to other
-fields of study, urban research deliberately avoided for the
first half of the twentieth century the methodology and
generalizations favored by the social sciences and remained

preoccupied with explaining the urban environment within the context of the national history. Social relations, so important to the founders of the great European schools of social thought, were not seen by American urban researchers as relevant to the uniqueness of the American development.

A. The Turner Thesis Restated

In the early years of this century urban research was scrutinized in the works of several writers affiliated with the University of Chicago. This so-called Chicago School,[1] whose proponents were sociologists, challenged the traditional approach to urban research which regarded urbanization only as a part of the national history. These urban sociologists insisted that American urbanization was a general process which was more easily understood as an ecological concept. This effort to change the focus was not then successful. Traditional urban life studies, biographies of individual cities, and economic development research dominated the literature of urban studies well into the 1950s.[2] In 1940, however, the course of urban research was effectively questioned. Arthur Schlesinger, Sr., through his article, "The City in American History,"[3] suggested that the urban environment was a significant topic for research around which other events in the U.S. national history could be organized. This apparent restatement of the so-called Turner Thesis,[4] which defined American society in terms of the constantly moving Western frontier but with an urban focus, was convincing enough for many urban researchers[5] to accept urbanization as the most significant event in the national history just as an earlier generation of social historians had accepted the westward expansion as that most significant event.

Expanding upon Schlesinger's suggested adoption of the hypothesis and method of Frederick Jackson Turner, several urban researchers were led to accept Turner's contention that population was the key to understanding the nature of American society. Although unable to support Schlesinger's intentional bonding of the urban environment with the U.S. national history in a cause-effect relationship, Eric Lampard, beginning in the 1950s,[6] wrote several important studies which defined urbanization in terms of population. Lampard's works reemphasized Turner's use of population movement to define the evolution of U.S. society as the framework most suitable for urban researchers to adopt.

B. Drift Towards Definition

The work of Arthur Schlesinger, Sr., stimulated the
attention of social researchers upon the city but his suggested direction for research efforts was controversial and
criticized by many. William Diamond[7] found Schlesinger's
article to be unempirical and its definition of urbanization
to be too inclusive in that all events or phenomena found
in the urban environment were considered significant. For
others, the earlier-discarded Turner Thesis was not worthy
of resurrection in whatever disguise. Throughout the two
decades following the Second World War attempts at providing
definition and focus for research into the urban environment
were numerous. The traditional methodology of the humanist
researchers and the method and generalizations favored by
social scientists alternated as the new approach to urban
studies during the period. Richard Wade's *The Urban Frontier*
(1959)[8] focused urban research in the early 1960s upon the
conventional approach to urbanization which stressed that the
urban environment was significant only as part of the larger
national history. Leo Schnore's work[9] attempted to reclaim
urban research for the Chicago School which emphasized
urbanization as a general process, not just a part of the
national history. Both Sam Warner[10] and Oscar Handlin[11]
tried to create a middle ground between the traditional historical view and the social science view in which urban research could exist. Their efforts enjoyed little success.
The most influential, though controversial, work during the
period was done by Eric Lampard.[12] Following the methodology
used by Turner and Schlesinger, Lampard insisted that a com-
prehensive theory of community had to be developed in order
to differentiate between what were urban phenomena worthy of
study and other events not relevant to urban research. Demographic analysis forms the major contribution Lampard made to
urban research. Population movement was the key to defining
urbanization according to Lampard. This analysis of cities
based upon quantified data such as census records was in-
sufficient, however, for most urban researchers. Roy Lubove[13]
was a forceful critic of Lampard's methods and conclusions.
Lubove found urbanization to be an abstraction not a concrete
process. Data analysis, according to Lubove, minimized the
important role that behavioral and subjective phenomena had as
agents for change within the urban environment.

C. The Quantification Revolt

In 1964, Stephen Thernstrom published his *Poverty and Progress: Social Mobility in a Nineteenth Century City*. This work, using already-developed methods of data analysis for demographic records, caused a reorientation in urban research which is only now being examined again. Thernstrom successfully linked hypothesis, data analysis, and computer technology into an attractively written work whose thrust was the examination of the lower strata of U.S. society: the common person within the urban environment. This work established the acceptability of the bottom-up focus for urban researchers. Upward social mobility so prominent in American folklore was in reality also an American political and social tenet. The use of computers to analyze vast quantities of census data was an innovation in research which had strong appeal to social scientists who were eagerly providing answers to society's questions during the 1960s. This revolution in urban research was quickly digested and given focus in the classrooms of American universities through use of textbooks such as *A History of Urban America* (1967) by Charles N. Glaab and A. Theodore Brown, and the *Urbanization of Modern America* (1973) by Zane Miller.

Other urban researchers refined the methodology connected to this quantification approach. The major methodological innovations associated with the quantification school involved the use of systematic sources of individual-level data such as manuscript census records, city directories, tax assessment lists, and other vital records to ascertain a set of common variables for a given population. These demographic characteristics were then linked to one another, coded, and statistically correlated. John Sharpless and Roy Shortridge analyzed bias in census manuscripts.[14] John Phillips worked on sampling and record linkage.[15] These efforts were important in advancing and refining the original methodology established in Thernstrom's work.

Theodore Hershberg, director of the Philadelphia Social History Project, expanded the quantification approach to urban research into an interdisciplinary focus. Demographers, sociologists, geographers, economists, and historians, according to Hershberg, must all collaborate in an effort to produce needed research into the process of urbanization. Hershberg's recent work on Philadelphia[16] attempts to provide a foundation for this view of urban research. This expansion of the quantification approach defines the interrelationship of human behavior, group experience, and the environment as the actual urban process. Critics of Hershberg's work contend that reliance on collaboration among types of urban researchers

is an ideal that will not occur in practice. Others find
Hershberg's concept of urban process unacceptable because it
ignores power relationships in urban society. Still others
see quantification as used by Hershberg to be a limited tool
which can become the end product of research and not just a
means to the better understanding of the urban environment.
This last criticism of Hershberg and those researchers
who follow his lead is the basic argument used by many who
totally reject the premises of the quantification revolution.
Full illumination of the facts, statistics, or whatever
figures which can be manipulated in census manuscripts, tax
roles, and city directories by computers can too easily become
the research project with interpretation of these figures only
vaguely given if at all by the researchers. This accountant's
research is the phenomenon rejected by critics of quantification. Sympathy for the real human lives which are represented
by any statistics uncovered is needed in order to attain the
essential understanding of the urban environment for many
engaged in the study of the city and its inhabitants. Quantification, many urban researchers believe, can provide a means
by which a more full account of the urban environment is
achieved but they do not see that quantification is synonymous
with urbanization as a concept.

D. Marxist Analysis Ignored

In 1962, William Appleman Williams published his *Tragedy
of American Diplomacy*. This work was a penetrating examination
of U.S. diplomatic history using the dialectical method favored
by historians trained in Marxist theory and analysis. In 1965,
Eugene D. Genovese used class analysis to discuss Afro-American
plantation life.[17] Both of these important contributions to
American historical literature were avowed attempts to give
Marxist analysis a place in interpreting the national history.
Until recently, Marxist methodology has not been represented
within the literature of urban studies. American researchers,
avoiding for decades the social theory developed in Europe,
also had learned from the Red Scare of the 1920s and the Mc-
Carthy methods in the 1950s to accept that Marxist interpretations and methods in social research do not promote publication
or career.[18] Today, however, some American social researchers
have begun to challenge the contention that Marxism is to be
seen only as a political force. Marxist methodology, for
some, can be divorced from its political movement and become
a useful intellectual tool. Marxist scholarship in Europe
has contributed much toward developing social theory which
can be useful in assimilating and interpreting data and con-

crete situations even if the techniques are also used to analyze class or other experiences as central aspects of social
life and change.

Since the mid-1970s several important contributions to
urban research have been made through use of the Marxist
methodology. In 1976 appeared Alan Dawley's *Class and Community: The Industrial Revolution in Lynn.* This excellent
study of the relationship between workers and the community
of Lynn provides useful insight into urban social change and
development during an important phase of urban growth. Other
works by Roy Rosenzweig,[19] Michael E. Stone,[20] and Betsy
Blackmar[21] have also used Marxist methods to explain urban
phenomena. The avoidance of Marxist analysis in social history by American urban researchers is a fact. But the efforts
of recent urban researchers to extract method from political
movement bring a certain respect for the possibility that dialectical method can be used to explain in a theoretical manner
the urban environment.

E. Continuing Research

With the exception of the few Marxist-oriented studies,
urban research has continued its indifference to social theory
and instead promoted a pluralism in methodology which gives
rise first to one school of "accepted" approach to research
then another. Recent research into the American urban environment has centered upon these themes: the relationship between
the federal government and municipalities,[22] the impact of
technology upon the urban environment,[23] and the growth of
cities in the so-called sunbelt.[24] These concentration points
for research have continued the fragmented approach to theory
and method which many scholars contend has so far provided
such great diversity and creative imagination within the literature of urban studies. In many respects, the original di-
vergence from the nineteenth-century approach to urban research
begun by the Chicago School in the 1920s has been sustained
in all the intervening years.[25] The Chicago School maintained
that there were specific themes in the American urbanization
process which related to the changing urban social and spatial
structure. Understanding the environment as an ecological
concept is still the only dominating thread connecting the
lengthy list of contributions which comprise the literature
of urban studies.

NOTES

1. For additional information about the Chicago sociologists, see Fred H. Matthews, *Quest for an American Sociology: Robert E. Park and the Chicago School* (Montreal, Canada: McGill-Queens University Press, 1977).

2. A model for urban biography was Bayrd Still's *Milwaukee: The History of a City* (Madison, WI: State Historical Society of Wisconsin, 1948).

3. Arthur M. Schlesinger, Sr. "The City in American History," *Mississippi Valley Historical Review* 27 (June 1940), 43-66.

4. Frederick Jackson Turner's work is reviewed in Richard M. Andrew's article "Some Implications of the *Annales* School and Its Methods for a Revision of Historical Writing about the United States," *Review* 1 (Winter/Spring 1978), 171-173 and in Ray A. Billington's *Frederick Jackson Turner: Historian, Scholar, Teacher* (New York: Oxford University Press, 1973).

5. Among the writers who followed Schlesinger's approach were Oscar Handlin, *Boston's Immigrants* (Cambridge, MA: Harvard University Press, 1941) and Carl Bridenbaugh, *Cities in Revolt* (New York: Alfred A. Knopf, 1955).

6. See for example, Eric Lampard, "The History of Cities in the Economically Advanced Areas," *Economic Development and Cultural Change* 3 (January 1955), 86-136.

7. William Diamond, "On the Dangers of an Urban Interpretation of History," in *Historiography and Urbanization*, edited by Eric Goldman (Baltimore, MD: Johns Hopkins University Press, 1941), pp. 67-108.

8. Richard C. Wade, *The Urban Frontier: The Rise of Western Cities, 1790-1830* (Cambridge, MA: Harvard University Press, 1959).

9. See for example, Leo Schnore, "On the Spatial Structure of Cities in the Two Americas," in *The Study of Urbanization*, edited by Philip Hauser and Leo Schnore (New York: Wiley, 1965), pp. 368-387.

10. Sam Bass Warner, Jr., *Streetcar Suburbs: The Process of Growth in Boston, 1870-1900* (Cambridge, MA: Harvard University Press, 1962).

11. Oscar Handlin, "The Modern City as a Field of Historical Study," in *The Historian and the City*, edited by Oscar Handlin and John Burchard (Cambridge, MA: MIT Press, 1963) pp. 1-26.

12. For example, see Eric Lampard, "Urbanization and Social Change: On Broadening the Scope and Relevance of Urban History," in *The Historian and the City*, edited by Oscar Handlin and John Burchard, pp. 225-247; and "American Historians and the Study of Urbanization," *American Historical Review* 67 (October 1961), 49-61.

13. Roy Lubove, "The Urbanization Process: An Approach to Historical Research," *Journal of the American Institute of Planners* 23 (January 1967), 33-39; and *Twentieth Century Pittsburgh: Government, Business, and Environmental Change* (New York: Wiley, 1969).

14. John B. Sharpless and Roy M. Shortridge, "Biased Under Enumeration in Census Manuscripts," *Journal of Urban History* 1 (August 1975), 409-439.

15. John A. Phillips, "Achieving a Critical Mass While Avoiding an Explosion: Letter Cluster Sampling and Normal Record Linkage," *Journal of Interdisciplinary History* 9 (Winter 1979), 493-508.

16. Theodore Hershberg, *Philadelphia: Work, Space, Family and Group Experience in the Nineteenth Century* (New York: Oxford University Press, 1981).

17. Eugene D. Genovese, *The Political Economy of Slavery* (New York: Pantheon, 1965); *The World the Slaveholders Made* (New York: Pantheon, 1969); *In Red and Black: Marxian Explorations in Southern and Afro-American History* (New York: Pantheon, 1971); and *Roll, Jordan, Roll: The World the Slaves Made* (New York: Pantheon, 1974).

18. A Marxist view of the urban environment is seen in the work of David Montgomery, "The Working Classes of the Pre-industrial American City, 1780-1830," *Labor History* 9 (Winter 1968), 3-22.

19. Roy Rosenzweig, "Middle-class Parks and Working-class Play: The Struggle over Recreational Space in Worcester, Massachusetts, 1870-1910," *Radical History Review* 21 (Fall 1979), 31-48.

20. Michael E. Stone, "The Housing Problem in the United States: Origins and Prospects," *Socialist Review* 10 (July/August 1980), 65-119.

21. Betsy Blackmar, "Rewalking the Walking City: Housing and Property Relations in New York City, 1780-1840," *Radical History Review* 21 (Fall 1979), 131-150.

22. See Mark Gelfand, *A Nation of Cities: The Federal Government and Urban America, 1933-1965* (New York: Oxford

University Press, 1975); and Philip Funigiello, *The Challenge to Urban Liberalism: Federal-City Relations During World War II* (Knoxville, TN: University of Tennessee Press, 1978).

23. See Mark Foster, *From Streetcar to Superhighway: American City Planners and Urban Transportation 1900-1940* (Philadelphia, PA: Temple University Press, 1981); Martin Melosi, *Garbage in the Cities: Refuse, Reform, and the Environment 1880-1980* (College Station, TX: Texas A & M University Press, 1981); and *Pollution and Reform in American Cities 1870-1930* (Austin, TX: University of Texas Press, 1980).

24. See Carl Abbott, *The New Urban America: Growth and Politics in Sunbelt Cities* (Chapel Hill, NC: University of North Carolina Press, 1981); and Richard M. Bernard and Bradley R. Rice, *Sunbelt Cities: Politics and Growth Since World War II* (Austin, TX: University of Texas Press, 1982).

25. The work of Richard Wade, Eric Lampard, and Leo Schnore in addition to recent studies of urban-related themes represent a continuation of the Chicago School which sought to define urbanization as an ecological concept with variations on a common theme.

GUIDE TO USE OF THIS BOOK

This bibliography is not all inclusive. Urban research, cross-disciplinary in nature, has links to the social, behavioral, historical, and the physical sciences. The number of publications appearing annually in all the possible related fields which conduct research into urban affairs is so cumbersome as to preclude any logical arrangement for such numbers or guarantee that there is reasonable access to useful and important items found within any such arrangement. Selective and representative are the best descriptions for the bibliography. There are over 2000 citations provided. It is possible using this sample group to begin research into most topics relevant to the urban environment.

Compilation of the bibliography was thus guided by a need to limit the total number of citations listed. Books and journal articles constitute the vast majority of citations provided. There are other forms of publications which do appear. Significantly, doctoral dissertations are included in the chapter on urban socialism. The amount of information available on this subject is small. A number of important dissertations on the subject remain unpublished. In order to provide a representative listing of works on the subject of urban socialism, dissertations were included. However, conference proceedings, working papers, and association studies are largely ignored within the bibliography. Furthermore, government publications have been included only when essential to representing the nature of research being conducted on a given topic or aspect of urban studies.

Recent research was a guideline for the citation-gathering process. As with all bibliographies, however, even as they are published significant works appear which date the total usefulness of the listing provided. For all bibliographic chapters but the one concerned with urban socialism, the years covered are 1973 to 1983, although a majority of items in each section do date from 1979 to 1982. Urban socialism was exempted from this time frame because research has been so scant in the subject that an expanded chronological base for publications was necessary in order to create a representative list of relevant research so far published. There are other

apparent exceptions to the chronology for publications but
these occur only for important studies needed to reflect ac-
curately the nature of the research efforts so far published.

English language is another limitation used for citations
listed in the bibliography. There is, nevertheless, a large
quantity of relevant studies published in foreign languages
which can be found through use of the various abstracts and
indexes which provide access to the literature of urban
studies. In the process of gathering citations there were
a number of significant foreign language studies of the
American urban environment which had to be discarded in order
to preserve a reasonably-sized bibliography. But, these works
should be sought out by the serious urban researcher in order
to gain a comparative analysis of common concerns for urban
life and activity.

The possible subdivisons of urban studies and the number
of topical interests represented in urban research are as
large as the number of subject disciplines which are involved
in the study of the urban environment. Arrangement of the
quantity of citations gathered for this bibliography is based
upon a subjective decision to provide representative research
on the urban environment of the several geographic divisions of
the United States and a desire to highlight certain topics
within urban research that are significant presently or that
have received little or no consideration in former urban-related bibliographies.

Structure of This Bibliography

The bibliography is divided into six parts. First there
is a general introduction to the literature of urban studies.
Then there is this guide followed by an annotated listing of
abstracts, indexes, and periodicals useful in compiling the
bibliography and important for urban research in general.
Thirdly the portion of citations relating to research conducted on the various geographic subdivisions of the United
States is presented. These citations are followed by six
chapters devoted to special topics relevant to research on
the urban environment. Finally, there are two indexes, geographic and topical, which provide additional access to the
items listed in the bibliographic chapters.

All citations have been assigned a number. These numbers
have been arranged sequentially starting with the United
States (General) chapter and continuing to the end of the
bibliography. Each citation contains the author's last name
and first initial. This abbreviation of author identity was
necessitated by the large number of authors who prefer using

initials which could not be traced to the actual first name.
This was particularly a problem with journal articles.
Rather than present a blend of first names and first name
initials, all author identities were abbreviated to the use
of initials. The remaining components of each citation follow standard form: title, place of publication, publisher,
and date of publication or copyright. Periodical citations
include title of article, title of periodical, volume number,
the issue number and/or month or quarter of issue, and pagina-
tion.

ABSTRACTS AND INDEXES

There are several services which provide indexing and abstracts for literature relating to the urban environment. In addition, using subject headings such as city, municipal, and urban will locate relevant information in those indexes and abstracts which cover subject fields in the social and behavioral sciences as well as history. The following titles of indexes and abstracts are representative of those providing the most useful access to recent research on the urban environment which has been published in books, journals, and conference proceedings.

Abstracts in Anthropology. Farmingdale, NY: Baywood Publishing Company, Inc., 1970- ; quarterly.

Abstracts for this listing are compiled from books, journal articles, and conference proceedings which have been written in the English language. A convenient directory of publishers and their addresses is provided for all abstracted items. There are separate author and subject indexes. Useful subject headings are rural-urban, social problems, urban anthropology, and urbanization.

America: History and Life: A Guide to Periodical Literature. Santa Barbara, CA: ABC-CLIO Press, 1964- ; quarterly.

The history of the United States and Canada is the focus of this guide to recent literature. There are four major sections to this publication. The most useful for urban research are parts A and B. Part A is a listing of abstracts and bibliographic citations from over 1500 U.S., Canadian, and European periodicals. It also contains abstracts and citations organized into geographic categories which are further subdivided chronologically. A subject index is provided for Part A. Useful subject headings are community, housing, urban renewal, urbanization, and welfare. Part B is a listing of book reviews accessed through either an author or subject index provided.

Education Index. Bronx, NY: H.W. Wilson Company, 1929- ;
   monthly.

   Focusing upon all fields of education, this index covers
not only periodicals but includes some monographs and specialized yearbooks as well. The arrangement is by subject and
author's last name. There is a book review section with
listings by author. Useful subject headings are cities and
towns, city planning, city schools, and urban education.

Housing and Planning References. Washington, DC: U.S. Department of Housing and Urban Development, 1953-1983; bimonthly.

   This index is a classified listing of publications relating
to concerns of local governments which have been received by
the library of the U.S. Department of Housing and Urban Development. Books, journal articles, pamphlets, and planning
reports are listed with a single-sentence annotation provided
for each. There are separate geographic and author indexes.
Useful subject headings are city growth, city planning,
municipal services, neighborhood conservation and preservation,
and urbanization.

Human Resources Abstracts. Ann Arbor, MI: University of Michigan. Institute of Labor and Industrial and Labor Relations,
   1966- ; quarterly.

   An interdisciplinary approach to urban literature, *Human
Resources Abstracts*, includes feature articles and analyses of
current urban problems as well as abstracts of books and
journal articles. Arrangement is by broad subject category
of which ethnic and minority group problems, immigration and
migration issues, and social services and welfare are of
interest for urban research. There is a separate author and
subject index.

Journal of Economic Literature. Pittsburgh, PA: American
   Economic Association, 1969- ; quarterly.

   Indexing all recent economic literature found in significant
periodicals, essays, government publications, and dissertations, the *Journal of Economic Literature* also provides abstracts for numerous items listed. Feature research articles
are included in each issue. There is a classified section
of book reviews. A listing of current periodicals by title
with a separate author index for authors listed in the section
is useful for urban research.

Psychological Abstracts. Lancaster, PA: American Psychological
   Association, Inc., 1927- ; monthly.

Providing international coverage of books, journal articles, and technical reports in psychology and related fields, *Psychological Abstracts* is arranged into sixteen major subject groups of which the environmental psychology and environmental issues subdivisions of the applied psychology section are the most useful for urban research. There is a separate subject index provided.

Public Affairs Information Service Bulletin. (PAIS). New York: Public Affairs Information Service, Inc., 1913- ; semimonthly.

A classified index to books, journal articles, pamphlets, reports, government publications, publications of significant societies and associations, and government legislation, *Public Affairs Information Service Bulletin* provides access to information on economics, social conditions and public affairs in general. There are a number of useful subject headings relating to the urban environment among which are city planning, municipal finance, municipal services, urban conditions, urban research, and urbanization.

Sage Urban Studies Abstracts. Beverly Hills, CA: Sage Publications, 1973- ; quarterly.

Approximately 250 abstracts of recent literature on the urban environment are compiled quarterly from books, journal articles, pamphlets, government publications, speeches, and legislative studies by the staff of *Sage Urban Studies Abstracts*. Arrangement is by general subject categories relating to urban studies such as housing and social services, urban history, and urban planning. There are separate author and subject indexes.

Social Sciences Citation Index (SSCI). Philadelphia, PA: Institute for Scientific Information, 1973- ; quarterly.

Providing access to recent literature in the social and behavioral sciences, *Social Sciences Citation Index* reviews a number of journals and books published world wide. There are four parts to the index: a Citation Index arranged by author, a Corporate Index for non-personal cited authors including their geographical location and important affiliations, a Source Index which lists who cited whom where, and a Subject Index. Useful citations for urban research can be found in the Subject Index by use of the term urban which produces the following entries: urban community, urban development, urban environment, urban growth, urban planning, and urban policy.

Social Sciences Index. New York: H.W. Wilson Company, 1974- ;
   quarterly.

   Covering the major periodicals in the social sciences,
this index provides a subject and author access to important
articles in anthropology, economics, environmental science,
geography, law planning and administration, and criminology,
all of which relate to research on the urban environment.
A book review section is also provided. Useful subject headings are cities and towns, municipal services, urban economics,
urban life, urban renewal, and urban transportation.

Sociological Abstracts. San Diego, CA: Sociological Abstracts,
   Inc., 1953- ; 5x yearly.

   Providing access to recent literature in sociology and related fields, *Sociological Abstracts* covers numerous journals,
monographs, and conference papers published throughout the
world. Arrangement is by broad subject categories relating
to the study of sociology. Within this arrangement, categories
of interest to urban research are urban structure and ecology,
social change and economic development, and social problems
and social welfare. There is both an author and subject
index. In addition, there is a source index provided for abstracts listed. Among the useful subject headings are cities,
municipalities, urban, urbanism, and urbanization.

Urban Affairs Abstracts. Washington, DC: National League of
   Cities and the U.S. Conference of Mayors, Library Services,
   1970- ; weekly.

   Periodical articles of general and specific relevance to
urban studies contained in several hundred journals, news-
letters, and other publications received by the National League
of Cities Library comprise the abstracts contained in *Urban
Affairs Abstracts.* Arrangement is by broad subject categories
of which community development, finance, housing, and municipal
administration are useful to urban research. There is a
Periodical Guide which lists information about periodicals
represented in the abstracts provided.

Other Abstracts and Indexes

Compendium of Research Reports   *International Bibliography of*
Criminal Justice Abstracts   *Political Science*
Environmental Abstracts   *Population Index*
Historical Abstracts   *Vertical File Index*
Humanities Index   *Women Studies Abstracts*
Index to Current Urban Documents

PERIODICALS

Urban research is a complicated and specialized activity and thus most periodicals in the field are aimed at a scholarly or practitioner audience. Several types of periodicals have been published in the field. There are periodicals that provide informational overviews of the urban environment. There are also journals that are theoretical in approach and rely on mathematical or economic models to explain urban issues. Further, there are magazines which are written for practicing urban planners and managers. Finally, there are special interest or single urban issue-oriented publications. The following annotated list of urban-related periodicals is representative of the various approaches and formats presently being published.

Computers, Environment, and Urban Systems. Elmsford, NY: Pergamon Press, 1976- ; quarterly.

This journal provides theory and opinion regarding the role of computers in planning human activities in the environment. Its primary focus is upon computer-based methodology as it is used to address specific problems confronting urban society. Subject coverage includes urban planning and urban policy in addition to computers and environmental science. Recent articles have addressed systems frameworks for environmental research planning and the preparation of a regional socioeconomic balance sheet.

International Journal of Urban and Regional Research. London: Edward Arnold Publications, 1977- ; quarterly.

Aiming for an international coverage of problems in urban and regional development as they relate to economic, social, and political institutions, this journal lists short abstracts in French, German, and Spanish for each feature article. In addition to the western world both developing and socialist nations are given consideration. There are short reports on current developments in urban policy and political events affecting urban affairs. Recent articles have examined housing, land use, and urban sociology.

Journal of Housing. Washington, DC.: National Association of
   Housing and Redevelopment Officials, 1941- ; bimonthly.

Directed at public officials, urban planners, and private
sector urban developers, the *Journal of Housing* gives focus
to the political, environmental, economic, and social conditions which affect housing for the low and moderate income
groups. The journal is divided into five news sections: News
Reports, Federal Reports, State Reports, Court Reports, and
City Reports. Recent highlights in these sections have provided information useful to research in community development
and neighborhood preservation.

Journal of the American Planning Association. Chicago, IL:
   American Planning Association, 1925- ; quarterly.

Addressing the activities of urban planners on the local,
state, and national levels, the journal's focus is upon
physical, economic, and social aspects of the urban environment which affect the work of urban planners. Special sections
include "Research Reports" and "Planners Notebook." There
are both a book review section and a section listing recent
periodical literature relating to urban affairs. Recent
articles have discussed urban growth and national urban
policy.

Journal of Urban Analysis and Public Management. London:
   Gordon and Breach Science Publications, 1972- ; semiannually.

This journal states as its goal the promotion and encourage-
ment of professional and innovative approaches to public sector
service delivery. Several interesting sections are common to
each issue: professional papers, working papers, communications,
and book reviews. The journal's primary focus is the presentation of papers which provide solutions to urban problems in
housing, municipal finance, and public services delivery.

Journal of Urban Economics. New York: Academic Press, 1974- ;
   bimonthly.

Research in urban economics and current economic problems
is the focus of this journal. All types of research studies
whether theoretical, empirical, positive, or normative are
reviewed for inclusion. There are usually six to eight feature
articles in each issue. Recent issues have analyzed federal
block grants, housing, public expenditures, urban growth, and
urban transportation.

Journal of Urban History. Beverly Hills, CA: Sage Publications, 1974- ; quarterly.

Urban history from all time periods and areas of the world
is the focus of this journal. Special attention is given
to new research techniques, new interpretations, or efforts
to compare societies over time spans or geographic areas.
There is a separate book review section. Recent articles have
addressed issues such as city planning, municipal health,
African cities, urban economy, and Richmond, Virginia.

Neighborhood: The Journal for City Preservation. New York:
   New York Urban Coalition, 1979- ; quarterly.

Though the focus of this journal is upon New York City
neighborhoods, its coverage of social, political, and economic
conditions of not only middle- to upper-class neighborhoods
but also the poor and declining areas of New York City attracts
a wide reading audience. Special effort is given to providing
in-depth interviews with residents of the various neighborhoods
covered in order to reveal the quality of urban life.

Planning. Chicago, IL: American Planning Association, 1972- ;
   monthly.

Planning is a journal of municipal management. There is a
monthly summary of important news affecting housing, land use,
transportation, and urban planning. Five feature articles
appear in each issue relating to a topic or single metropolitan
area. Recent articles have examined Austin, Texas, ethics in
urban planning, and city-state relations.

Urban Affairs Quarterly. Detroit, MI: University of Michigan.
   Institute of Public Policy Studies, 1965- ; quarterly.

Urban Affairs Quarterly is intended to facilitate "an inter-
change of ideas and concerns between those engaged in basic
or applied urban research and those responsible for making
or implementing public policy and programs." A primary con-
cern is to present the impact of policy decisions on those
who live in urban areas. Recent articles have scrutinized
the problems of neighborhood preservation, housing policy,
and the urban political process.

Urban Anthropology. Brockport, NY: SUNY-Brockport College,
   Department of Anthropology, 1972- ; quarterly.

Urban Anthropology is produced by the Institute for the
Study of Man. Though its primary focus is upon the cultural
systems within cities, urbanization, rural migration to urban
areas, housing, and ethnic groups are also frequent topics of
feature articles. There is a predominance of articles
written by anthropologists, but urban research by other

social scientists does occasionally appear. Recent articles have addressed the questions of urban politics and urban ethnic group experiences.

Urban Land. Washington, DC: Urban Land Institute, 1941- ; monthly.

Urban land use is the focus of this journal. Several unique features of the journal which relate to land development and planning are the "Solution File," "Regulatory Review," and "Environmental Comment." There is a book review section included. Three or four feature articles concerned with a single issue in urban land use appear each issue. Recent articles have discussed transportation for growing urban areas and government-sponsored economic development initiatives.

Urban Studies. Edinburgh, UK: University of Glasgow, 1964- ; 4x year.

Urban Studies is oriented primarily toward British urban problems, but there is usually one or more feature articles which address American urban problems each issue. Although emphasis within the journal is upon urban economics, housing, demography, and planning are also topics. There are separate book reviews and books received sections. Recent articles have addressed the issues of fiscal conditions of American cities and the urban renewal activities of several American cities.

Urbanism Past and Present. Milwaukee, WI: University of Wisconsin-Milwaukee, Department of History, 1975- ; semiannual.

The journal is intended as a way of exchanging theories and techniques for discovering the process by which cities develop. The functioning of urban institutions is a frequent topic. Important information concerning urban research, conferences, and activities of associations appears in the "Peripatetic Urbanist" section. There is an extensive bibliography of recent books and journal articles relating to urban studies in each issue. Recent articles have addressed such topics as cities in Oregon and suggested planning goals for the city of Milwaukee.

Other Periodicals of Interest

Urban Studies Focus

American City and County
American Preservation
American Real Estate and
  Urban Economics Association
  Journal
Computers, Environment and
  Urban Systems
Education and Urban Society
Ekistics
Environment and Behavior
Environment and Planning A:
  International Journal of
  Urban and Regional Research
Historic Preservation
International Journal of
  Urban and Regional Research
Journal of Regional Science
Journal of the American
  Planning Association
Journal of Urban Analysis
  and Public Management
Journal of Urban and Environ-
  mental Affairs
Journal of Urban Economics
Journal of Urban History

Journal of Urban Law
  (University of Detroit)
Landscape
Metro
National Civic Review
Nation's Cities
Planning
Planning and Administration
Practicing Planner
Regional Science and Urban
  Economics
Socio-economic Planning
  Sciences
South Atlantic Urban
  Studies
Urban and Social Change
  Review
Urban Ecology
Urban Education
Urban Forum/Colloque Urbain
Urban Interest
Urban Land
Urban Life
Urban Review
Urbanism Past and Present

Periodicals Related to Urban Studies

Afro-American New York Life
  and History
American Studies
Annals of Regional Science
Atlantic Historical Quarterly
Chicago History
Cincinnati Historical Society
  Bulletin
Demography
Detroit in Perspective
Environment
Ethnic and Racial Studies
Ethnicity

Ethnohistory
Historical Messenger (Mil-
  waukee)
Journal of American Studies
Journal of Black Studies
Journal of Ethnic History
Journal of Ethnic Studies
Journal of German-American
  Studies
Land Economics
Milwaukee History
New York History
Rochester History

Includes Topics Related to Urban Studies

Amerasia Journal
American Academy of Political
  Social Science Annals
American Behavioral Scientist
American Economic Review
American Education
American Historical Review
American Jewish Archives
American Jewish Historical
  Quarterly
American Journal of Economic
  Sociology
American Journal of Economics
  and Sociology
American Journal of Education
American Journal of Sociology
American Political Science
  Review
American Sociological Review
Annals of the Association of
  American Geographers
Anthropological Quarterly
Archaeology
Architectural Record
Aztlan
Catholic Historical Review
Challenge
Comparative Education
Crisis
Current Anthropology
Economic Geography
Economic Inquiry
Essays in Public Works His-
  tory
Feminist Studies
Focus
Geographical Bulletin
Geographical Review
Geography
Historian
Historical Methods News-
  letter
History
History of Education Quar-
  terly
Human Organization

Immigration History News-
  letter
International Migration Review
Jewish Journal of Sociology
Journal of American History
Journal of Applied Behavior
  Science
Journal of Economic History
Journal of Economic Issues
Journal of Family History
Journal of General Education
Journal of Historical Geography
Journal of Human Resources
Journal of Interdisciplinary
  History
Journal of Negro Education
Journal of Negro History
Journal of Political Economy
Journal of Political Science
Journal of Popular Culture
Journal of Social History
Journal of Social Psychology
Labor History
Law and Society Review
Mass Transit
Phylon
Policy and Politics
Polish American Studies
Political Science Quarterly
Professional Geographer
Prologue
Public Historian
Public Interest
Publius
Review of Radical Political
  Economics
Social Forces
Social Policy
Social Problems
Sociological Analysis
Sociological Quarterly
Transportation Journal
Transportation Quarterly

Occasional Topics Related to Urban Studies

Agricultural History
Alabama Historical Quarterly
Alabama Review
American Quarterly
American Review of Reviews
Annals of Iowa
Annals of Wyoming
Arizona and the West
Arkansas Historical Quarterly
Australian Journal of Politics and History
Business History Review
California History
Chronicle of Oklahoma
Church History
Cinema Journal
Civil War History
Commentary
Concordia Historical Institute Quarterly
Connecticut Historical Society Bulletin
Connecticut Review
Essex Institute Historical Collection
Filson Club Historical Quarterly
Florida Historical Quarterly
Georgia Historical Quarterly
Great Lakes Review
High School Journal
History of New Hampshire
Idaho Yesterdays
Indiana Magazine of History
Journal of Arizona History
Journal of Mississippi History
Journal of Religious History
Journal of Southern History
Journal of the Illinois State Historical Society
Journal of the West
Kansas History
Louisiana History
Louisiana Studies
Maryland Historical Magazine

Michigan Academician
Michigan History
Mid-America
Midwest Quarterly
Minnesota History
Missouri Historical Review
Nebraska History
New England Quarterly
New Jersey History
New Mexico Historical Review
New-York Historical Society Quarterly
Niagara Frontier
North Carolina Historical Review
Northwest Ohio Quarterly
Ohio History
Old Northwest
Oral History Review
Oregon Historical Quarterly
Pacific Historical Review
Pacific Northwest Quarterly
Palimpsest
Pennsylvania History
Pennsylvania Magazine of History and Biography
Plains Anthropologist
Polity
Register of the Kentucky Historical Society
Rhode Island History
Social Science History
Social Science Journal
Social Science Quarterly
Social Service Review
Socialist Review
Societas--A Review of Social History
South Carolina Historical Magazine
Southwestern Historical Quarterly
Southwestern Studies
Tennessee Historical Quarterly
Utah Historical Quarterly

Vermont History
Virginia Magazine of History
  and Biography
Western Historical Quarterly

Western Political Quarterly
William and Mary Quarterly
Wisconsin Magazine of History

GEOGRAPHIC QUINTET

Urbanization in the United States has been a twentieth-century phenomenon. Since 1940, however, the central feature of American urban growth has been visible in a migration of population away from the northern industrial centers of the nation toward the so-called sunbelt cities. This population shift has been detailed in *The New Urban America: Growth and Politics in Sunbelt Cities* (1981) by Carl Abbott. Abbott examined sunbelt area population growth, metropolitan expansion, rise in per capita income, and employment statistics. From this examination, Abbott concluded that there were two distinct sunbelt growth regions: one comprised of seven states located in the southeast, including Delaware, Maryland, and Virginia, the other made up of ten states and containing Washington, Oregon, Utah, and Colorado which are not always listed as sunbelt members.

Abbott's primary focus, however, was on the patterns of urban-suburban growth in the sunbelt as they affected urban and suburban politics. He stressed that the mobilization of United States resources to fight World War II was a critical factor in promoting the original growth patterns in the sunbelt cities. Population migration from the north to the sunbelt followed the awarding of federal government defense contracts to southern and southwestern locations. According to Abbott, this population shift toward the sunbelt was accelerated after 1945 because of the region's steady economic growth due to the location of defense and high technology industries in the sunbelt as well as the creation of activities associated with tourism, recreation, and retirement living which attracted northern population groups southward. The work is an important attempt to define and clarify recent conclusions reached from census data that the American urban populace is located increasingly in the south and southwest regions of the nation. An earlier but less influential work, *The Rise of the Sunbelt Cities* (1974) by David C. Perry and Alfred J. Watkins, attempted to outline the same phenomenon. Abbott's work has been followed by Raymond A. Mohl's *Sunbelt Cities: Politics and Growth Since World War II* (1982). These three works form the signifi-

cant research efforts conducted to date into the urban environment of the southern and southwestern regions of the United
States.

In the East, the work of Theodore Hershberg, *Philadelphia:
Work, Space, Family and Group Experience in the Nineteenth
Century* (1981) is the major study so far of the urbanization
process. Hershberg's study of Philadelphia is based on census
data and concentrates on the years 1850 to 1880. Three important elements are emphasized in this study: environment,
behavior, and group experience. The interrelationship between
these elements explains, according to Hershberg, the uniqueness
of American urban life. *The Divided Metropolis: Social and
Spatial Dimensions of Philadelphia, 1800-1975* (1980) by William
Cutler II and Howard Gillette, Jr., addresses key questions of
urban policy in Philadelphia. Regional consideration of urban
growth in the East is examined in *The Boston Region, 1810-1850:
A Study in Urbanization* (1980) by Francis Blouin. Though less
academic than the previous studies, several brief research
booklets have been published by the Brooklyn Educational and
Cultural Alliance: *The Shaping of a City: A Brief History of
Brooklyn* (1979) by David Ment and *Building Blocks of Brooklyn:
A Study of Urban Growth* by David Ment and David Framberger.
These publications provide starting points for further research
into the various boroughs of New York City.

Midwest cities have received considerable attention from
urban researchers. Milwaukee, St. Louis, and Chicago are all
frequent subjects of general and specific research on urban
concerns. Recently urban researchers have also examined less
important urban areas of the Midwest. Robert G. Barrows provided an excellent mobility study of Indianapolis, "Hurryin'
Hoosiers and the American Pattern: Geographic Mobility in Indianapolis and Urban North America" (*Social Science History*
5, 2 (1981): 197-222). Zane Miller's *Suburb: Neighborhood and
Community in Forest Park, Ohio 1935-1976* (1981) details the
growth of a middle-class suburb and concludes with the view-
point that Forest Park's development was similar to countless
other cities which sprang up in the Midwest after the Second
World War. Miller's opinions are confirmed in Frederick
Steiner's *The Politics of New Town Planning: The Newfields,
Ohio Story* (1981).

The five following chapters list books and journal articles
which have been published during the last ten years and represent research conducted into the urban environment of the
United States in general and specifically of the geographic
divisions of the nation: East, South, Midwest, and West.

UNITED STATES
(GENERAL)

1. Abrahamson, M., and M. DuBick. "National Dominance and Urban Exploitation." *Urban Affairs Quarterly* 15, 2 (1979): 146-163.

2. "Accommodating Diversity: Civic Leadership in Transition." *National Civic Review* 68, 11 (1979): 583-599.

3. Ahlgren, D. "Architectural Drawings: Sources for Urban History." *Urban History Review* 11, 3 (1983): 67-72.

4. Allen, W.R. "Class, Culture, and Family Organization: The Effects of Class and Race on Family Structure in Urban America." *Journal of Comparative Family Studies* 10, 3 (1979): 301-314.

5. Anderson, J., et al. *Redundant Spaces in Cities and Regions? Studies in Industrial Decline and Social Change.* New York: Academic Press, 1983.

6. Angle, J. "The Ecology of Language Maintenance: Data from Nine U.S. Metropolitan Areas." *Urban Affairs Quarterly* 17, 2 (1981): 219-232.

7. Angotti, T. "The Planning of the Open Air Museum and Teaching Urban History." *Museum* 34, 3 (1982): 179-188.

8. Appelbaum, R.P. *Size, Growth, and U.S. Cities.* New York: Praeger, 1978.

9. Arnold, J.L. "Suburban Growth and Municipal Annexation, 1745-1918." *Maryland Historical Magazine* 73, 2 (1978): 109-128.

10. Ashford, D.E. *National Resources and Urban Policy.* New York: Methuen, 1980.

11. Atlas, J., and P. Dreier. "The Housing Crisis and the Tenants Revolt." *Social Policy* 10, 4 (1980): 13-24.

12. Ayeni, B. "Intraurban Residential Migration: An Entropy Maximizing Approach." *Journal of Regional Science*, 19, 3 (1979): 331-344.

13. Bahl, R.W., et al. *Public Employment and State and Local Government Finance*. Cambridge, MA: Ballinger Publishing Co., 1980.

14. Baldassare, M. *The Growth Dilemma: Resident's Views and Local Population Change in the United States*. Berkeley, CA: University of California Press, 1981.

15. ———. *Residential Crowding in Urban America*. Berkeley, CA: University of California Press, 1979.

16. Ball, R.M., et al. "Revitalizing Older Cities." *National Civic Review* 67, 5 (1978): 228-234.

17. Barabba, V.P. "The Demographic Future of the Cities." *Current Municipal Problems* 7, 2 (1980): 180-190.

18. Barnes, W., and R. Abraham. "Pennsylvania Avenue—Main Street, America Comes to Life." *Urban Land* 38, 6 (1979): 12-17.

19. Barrows, R.G. "Beyond the Tenement: Patterns of American Urban Housing, 1870-1930." *Journal of Urban History* 9, 4 (1983): 395-420.

20. Barth, G. *City People: The Rise of Modern City Culture in Nineteenth-Century America*. New York: Oxford University Press, 1980.

21. Bayor, R.H. *Neighborhoods in Urban America*. Port Washington, NY: Kennikat Press, 1982.

22. Beale, C.L. "The Population Turnaround in Rural and Small Towns in America." *Policy Studies Review* 2, 1 (1980): 43-54.

23. Beaton, W.P., and L.B. Sossamon. "Housing Integration and Rent Supplements to Existing Housing." *Professional Geographer* 34, 2 (1982): 147-155.

24. Bedford, H.F. *Trouble Downtown: The Local Context of Twentieth-Century America*. New York: Harcourt Brace Jovanovich, 1978.

25. Bender, T. *Toward an Urban Vision: Ideas and Institutions in Nineteenth-Century America*. Baltimore, MD: Johns Hopkins University Press, 1982.

26. Blair, J.P., and D. Nachmias. *Fiscal Retrenchment and Urban Policy*. Beverly Hills, CA: Sage Publications, 1979.

27. Boyer, M.C. *Dreaming the Rational City: The Myth of American City Planning*. Cambridge, MA: MIT Press, 1983.

28. Boyer, P. *Urban Masses and Moral Order in America, 1820-1920.* Cambridge, MA: Harvard University Press, 1978.

29. Boyle, H.C. "Region vs. the Neighborhoods." *Social Policy* 12, 4 (1982): 3-9.

30. Boyle, J., and D. Jacobs. "The Intra-City Distribution of Services: A Multivariate Analysis." *American Political Science Review* 76, 2 (1982): 371-379.

31. Brademas, J. "Federal Reorganization and Its Likely Impacts on State and Local Government." *Publius* 8, 2 (1978): 25-38.

32. Brandon, D. "Suburban Renewal: Public Sponsored Industrial Development." *Urban Land* 41, 6 (1982): 15-19.

33. Brecher, C., and R.D. Horton. *Setting Municipal Priorities.* New York: Basic Books, 1981.

34. Brecher, J., et al. *Brass Valley: The Story of Working People's Lives and Struggles in an American Industrial Region.* Philadelphia, PA: Temple University Press, 1983.

35. Brown, L.B. "Main Streets Get Street Wise." *Historic Preservation* 21, 1 (1979): 29-34.

36. Brunn, S.D., and J.O. Wheeler. *The American Metropolitan System: Present and Future.* Somerset, NJ: John Wiley and Sons, 1980.

37. Bryce, H.J. *Planning Smaller Cities.* Lexington, MA: D.C. Heath, 1979.

38. Buczko, W. "Measures of Inequality, Distribution Standards, and Municipal Expenditures." *Social Science Quarterly* 63, 4 (1982): 661-673.

39. Burchell, R.W., and D. Listokin. *Cities Under Stress: The Fiscal Crises of Urban America.* Piscataway, NJ: Center for Urban Policy Research, 1981.

40. Burchell, R., and D. Listokin. *The Fiscal Impact Handbook: Estimating Local Costs and Revenues of Land Development.* New Brunswick, NJ: The Center for Urban Policy Research, 1978.

41. Burns, E., and R.W. Travis. "Small Town Growth and Metropolitan Community Evidence from United States Daily Urban Systems." *The Annals of Regional Science* 16, 1 (1982): 75-79.

42. Calhoun, D. "From Collinearity to Structure: San Francisco and Pittsburgh, 1860." *Historical Methods* 14, 3 (1981): 107-122.

43. Callow, A.B. *American Urban History: An Interpretive Reader with Commentaries.* New York: Oxford University Press, 1982.

44. Canuto, E., and A. Villa. "On Planning Dissaturation Control Signals for Urban Areas." *Transportation Research* 17B, 1 (1983): 45-54.

45. Caplow, T., and B. Chadwick. "Inequality and Lifestyles in Middletown, 1920-1978." *Social Science Quarterly* 60, 3 (1979): 367-386.

46. Caraley, D. "Congressional Politics and Urban Aid: A 1978 Postscript." *Political Science Quarterly* 93, 3 (1978): 411-420.

47. Cebula, R.J. "A Note on the Impact of Right-to-Work Laws on the Cost of Living in the United States." *Urban Studies* 19, 2 (1982): 193-197.

48. Checkoway, B. "Large Builders, Federal Housing Programmers, and Postwar Suburbanization." *International Journal of Urban and Regional Research* 4, 1 (1980): 21-45.

49. Chinitz, B. *Central City Economic Development.* Lanham, MD: University Press of America, 1984.

50. Christenson, J.A., and J.W. Robinson. *Community Development in America.* Ames, IA: Iowa State University Press, 1980.

51. ————. "Urbanism and Community Sentiment: Extending With's Model." *Social Science Quarterly* 60, 3 (1979): 387-400.

52. Chudacoff, H.P. *The Evolution of American Urban Society.* Englewood Cliffs, NJ: Prentice-Hall, Inc., 1981.

53. Clark, T.A. "Federal Initiatives Promoting the Dispersal of Low-Income Housing in Suburbs." *Professional Geographer* 34, 2 (1982): 136-146.

54. Clay, P.L. *Neighborhood Renewal: Middle-Class Resettlement and Incumbent Upgrading in American Neighborhoods.* Lexington, MA: Lexington Books, 1979.

55. Collver, A., and M. Semyonov. "Suburban Change and Persistence." *American Sociological Review* 44, 3 (1979): 480-486.

56. Cooper, J.L. *The Police and the Ghetto.* Port Washington, NY: Kennikat Press, 1979.

57. Cox, D.R., and F.Z. Nartowicz. "Jurisdictional Fragmentation in the American Metropolis: Alternative Perspec-

tives." *International Journal of Urban and Regional Research* 4, 2 (1980): 196-211.

58. Crider, G.I. "William Dean Howells and the Antiurban Tradition: A Reconsideration." *American Studies* 19, 1 (1978): 55-64.

59. Crimmins, T.J. "Past in the Present: An Urban Historian's Agenda for Public Housing and Historic Preservation." *Georgia Historical Quarterly* 63, 1 (1979): 53-60.

60. Danbom, D.B. *The Resisted Revolution: Urban America and the Industrialization of Agriculture, 1900-1930.* Ames, IA: Iowa State University Press, 1979.

61. Davies, R.L., and A.G. Champion. *The Future for the City Centre.* New York: Academic Press, 1983.

62. Dear, M., and A.J. Scott. *Urbanization and Urban Planning in Capitalist Society.* New York: Methuen, 1981.

63. Dommel, P.R., et al. *Decentralizing Urban Policy: Care Studies in Community Development.* Washington, DC: Brookings Institution, 1982.

64. Doucet, M.J. "Urban Land Development in Nineteenth-Century North America: Themes in the Literature." *Journal of Urban History* 8, 3 (1982): 299-342.

65. Dye, T.R., and J.H. Ammons. "Frostbelt and Sunbelt Cities: What Difference It Makes." *Urban Interest* 2, 1 (1980): 28-33.

66. Ecker-Racz, L. "State/Local Financial Crises: With Benefit of Hindsight." *National Civic Review* 68, 11 (1979): 605-612.

67. Eckert, J.K. "Urban Renewal and Redevelopment High Risk for the Marginally Subsistent Elderly." *Gerontologist* 19, 5 (1979): 496-502.

68. Fainstein, S.S. "American Policy for Housing and Community Development: A Comparative Examination." *Policy Studies Journal* 8, 2 (1979): 323-336.

69. Fleming, R.L. "Recapturing History: A Plan for Gritty Cities." *Landscape* 25, 1 (1981): 20-27.

70. Fosler, R.S., and R.A. Berger. *Public-Private Partnership in American Cities.* Lexington, MA: Lexington Books, 1982.

71. Foster, H.S., and B.R. Beattie. "Urban Residential Demand for Water in the United States." *Land Economics* 55, 1 (1979): 43-58.

72. Frieden, B.J. "Allocating the Public Service Cost of
     New Housing." *Urban Land* 39, 1 (1980): 12-16.

73. Fungigiello, J. *The Challenge of Urban Liberalism:
     Federal-City Relations during World War II.* Knoxville,
     TN: University of Tennessee Press, 1978.

74. Gale, D.E. "Middle-Class Resettlement in Older Urban
     Neighborhoods: The Evidence and the Implications."
     Journal of the American Planning Association 45, 3
     (1979): 293-304.

75. Gale, S., and E. Moore. *The Manipulated City: Perspec-
     tives on Spatial Structure and Social Issues in Urban
     America.* New York: Methuen, 1980.

76. Garber, S.R. "The Main Street Project: Downtown Economic
     Development within the Context of Historic Preservation."
     Challenge 10, 7 (1979): 12-19.

77. Gardner, D.S. "American Urban History: Power, Society
     and Artifact." *Trends in History* 2, 1 (1981): 49-78.

78. Gillard, Q. "Reverse Community and the Inner City Low-
     Income Problem." *Growth and Change* 10, 3 (1979): 12-18.

79. Gillette, H. "The Evolution of Neighborhood Planning:
     From the Progressive Era to the 1949 Housing Act."
     Journal of Urban History 9, 4 (1983): 421-444.

80. Gluck, P., and R.J. Meister. *Cities in Transition:
     Social Changes and Institutional Responses to Urban
     Development.* New York: New Viewpoints, 1979.

81. Goldin, C. "Household and Market Production of Families
     in a Late Nineteenth-Century American City." *Explora-
     tion in Economic History* 16, 2 (1979): 111-131.

82. Goodman, R. *The Last Entrepreneurs: America's Regional
     Wars for Jobs and Dollars.* New York: Simon and Schuster,
     1979.

83. Graham, H.D., and T.R. Gurr. *Violence in America: His-
     torical and Comparative Perspectives.* Beverly Hills,
     CA: Sage Publications, 1979.

84. Green, R.L. *The Urban Challenge--Poverty and Race.*
     Chicago, IL: Follett Publishing, 1977.

85. Greenberg, M.R. "A Note on the Changing Geography of
     Cancer Mortality Within Metropolitan Regions of the
     United States." *Demography* 18, 3 (1981): 411-420.

86. Greenberg, R.D. "Municipal Bankruptcy." *The Urban
     Lawyer* 10, 2 (1978): 266-288.

87. Greer, S., et al. *Accountability in Urban Society: Pub-
     lic Agencies Under Fire*. Beverly Hills, CA: Sage
     Publications, 1978.

88. Grieson, R.W. *The Urban Economy and Housing*. Lexington,
     MA: Lexington Books, 1983.

89. Grob, G.N. *The Mentally Ill in Urban America: Four
     Studies 1914-1922*. New York: Arno Press, 1979. Re-
     print.

90. Haugh, J.B. *Power and Influence in a Southern City:
     Compared with the Classic Community Power Studies of
     the Lynds, Hunter, Vidich, and Bensman*. Lanham, MD:
     University Press of America, 1980.

91. Hayden, D. *The Grand Domestic Revolution: A History of
     Feminist Designs for American Homes, Neighborhoods and
     Cities*. Cambridge, MA: MIT Press, 1982.

92. Henig, J.R. "Gentrification and Displacement Within
     Cities: A Comparative Analysis." *Social Science Quar-
     terly* 61, 3/4 (1980): 638-652.

93. Herman, R, et al. *Counsel for the Poor: Criminal Defense
     in Urban America*. Lexington, MA: D.C. Heath, 1977.

94. Higgs, R. "Cycles and Trends of Mortality in 18 Large
     American Cities, 1871-1900." *Explorations in Economic
     History* 16, 4 (1979): 381-408.

95. Hodge, D.C. *Inner City Revitalization and Displacement:
     The New Urban Future*. Seattle, WA: University of Wash-
     ington Press, 1979.

96. Hollingsworth, J.R., and E.J. Hollingsworth. *Dimensions
     in Urban History: Historical and Social Science Per-
     spectives on Middle-Size American Cities*. Madison, WI:
     University of Wisconsin Press, 1979.

97. Hosmer, C.B. *Preservation Comes of Age: From Williams-
     burg to the National Trust, 1926-1949*. Washington,
     DC: The Preservation Press, 1981.

98. Houseman, G.L. *City of the Right: Urban Applications of
     American Conservative Thought*. Westport, CT: Greenwood
     Press, 1982.

99. Houston, L.O. "Urban Policy: What Else Is New?" *Urban
     Land* 37, 7 (1978): 4-8.

100. Howett, M. "Frank Lloyd Wright and American Residential
     Landscaping." *Landscape* 26, 1 (1982): 33-44.

101. Hudnut, W.H. "The Federal System in the 1980's: A City

Perspective." *School of Public and Environmental Affairs Review* 3, 2 (1982): 27-30.

102. Hudson, W.E. "The Federal Aid Crutch: How a Sunbelt City Comes to Depend on Federal Revenue." *Urban Interest* 2, 1 (1980): 34-44.

103. Huth, M.J. "New Hope for the Revival of America's Central Cities." *Annals of the American Academy of Political and Social Science* 45, 1 (September 1980): 118-129.

104. Innes de Neufville, J. *The Land Use Policy Debate in the United States*. New York: Plenum Press, 1981.

105. Isserman, A.M., and K.L. Majors. "General Revenue Sharing: Federal Incentives to Change Local Government?" *Journal of the American Institute of Planners* 44, 3 (1978): 317-327.

106. Jackson, G., et al. *Regional Diversity: Growth in the United States, 1960-1980*. Boston, MA: Auburn House, 1981.

107. Jacob, H. *Crime and Justice in Urban America*. Englewood Cliffs, NJ: Prentice-Hall, Inc., 1980.

108. Jacobs, J. "DARE to Struggle: Organizing Urban America." *Socialist Review* 12, 3-4 (1982): 85-104.

109. Jaher, F.C. *The Urban Establishment: Upper Strata in Boston, New York, Charleston, Chicago and Los Angeles*. Urbana, IL: University of Illinois Press, 1981.

110. Jakubs, J.F. "Low-Cost Housing: Spatial Deconcentration and Community Change." *Professional Geographer* 34, 4 (1982): 156-166.

111. Jaye, M.C., and A.C. Watts. *Literature and the American Urban Experience*. Manchester: Manchester University Press, 1981.

112. Johnson, D.R. *Policing the Urban Underworld: The Impact of Crime on the Development of the American Police, 1800-1887*. Philadelphia, PA: Temple University Press, 1979.

113. Johnson, J.H., and C.G. Pooley. *The Structure of Nineteenth-Century Cities*. London: Croom Helm, 1982.

114. Johnston, R.J. *The American Urban System: A Geographical Perspective*. New York: St. Martin's Press, 1982.

115. Kammann, R., et al. "Unhelpful Behavior in the Street: City Size or Immediate Pedestrian Density?" *Environment and Behavior* 11, 2 (1979): 245-250.

116. Katz, M.B. "Social Class in North American Urban History." *Journal of Interdisciplinary History* 11, 4 (1980): 579-606.

117. Katzman, D. *Seven Days a Week: Women and Domestic Service in Industrializing America.* New York: Oxford University Press, 1979.

118. Kazis, R., and P. Sabonis. "CETA and the Private Sector Imperative." *Social Policy* 10, 4 (1980): 6-12.

119. Kearney, R.C. "The Unions and the Cities, or Barking Dogs Don't Bite." *Current Municipal Problems* 6, 2 (1979): 217-227.

120. Kettl, D.F. "Can the Cities Be Trusted? The Community Development Experience." *Political Science Quarterly* 94, 3 (1979): 437-452.

121. Krase, J. *Self and Community in the City.* Lanham, MD: University Press of America, 1982.

122. Krueckeberg, D.A. *The American Planner: Biographies and Recollections.* New York: Methuen, 1983.

123. ————. *Introduction to Planning History in the U.S.* Piscataway, NJ: Center for Urban Policy Research, 1982.

124. LaGory, M., et al. "The Age of Segregation Process: Explanations for American Cities." *Urban Affairs Quarterly* 16, 1 (1980): 59-80.

125. Lang, M.H. *Gentrification Amid Urban Decline: Strategies for America's Older Cities.* Cambridge, MA: Ballinger Publishing Co., 1982.

126. Laurie, Ian C. *Nature in Cities: The Natural Environment in the Design and Development of Urban Green Space.* New York: John Wiley and Sons, 1978.

127. Lee, B.A. "The Disappearance of Skid Row: Some Ecological Evidence." *Urban Affairs Quarterly* 16, 1 (1980): 81-107.

128. Listokin, D. *Landmarks, Preservation, and the Property Tax.* New Brunswick, NJ: Rutgers University Press, 1982.

129. Lloyd, W.J. "Understanding Late Nineteenth-Century American Cities." *Geographical Review* 71, 4 (1981): 460-471.

130. London, B. *The Revitalization of Inner City Neighborhoods: A Preliminary Bibliography.* Chicago, IL: Council of Planning Librarians, 1978.

131. Long, L., and D. DeAre. "The Slowing of Urbanization
     in the U.S." *Scientific American* 249, 1 (1983):
     33-41.

132. Longbrake, D., and J.F. Geyler. "Commercial Develop-
     ment in Small Isolated Energy Impacted Communities."
     Social Science Journal 16, 2 (1979): 51-62.

133. Lynch, J. "City Guides Aid in Preservation." *Challenge*
     11, 2 (1980): 4-9.

134. Machor, J.I. "The Garden City in America: Crevecoeur's
     Letters and the Urban-Pastoral Context." *American
     Studies* 23, 1 (1982): 69-84.

135. Mangum, G., and S.F. Seninger. *Coming of Age in the
     Ghetto: A Dilemma of Youth Unemployment*. Baltimore,
     MD: Johns Hopkins University Press, 1978.

136. Marcuse, P. "Housing in Early City Planning." *Journal
     of Urban History* 6, 2 (1980): 153-176.

137. Marshall, D.R. *Urban Policy Making*. Beverly Hills,
     CA: Sage Publications, 1979.

138. Marshall, H. "White Movement to the Suburbs." *American
     Sociological Review* 44, 6 (1979): 975-994.

139. Mason, J.B. *History of Housing in the U.S., 1930-1980*.
     Houston, TX: Gulf Publishing Co., 1982.

140. Massey, G., and D. Lewis. "Energy Development and
     Mobile Home Living: The Myth of Suburbia Revisited?"
     Social Science Journal 16, 2 (1979): 81-92.

141. Matlach, C. "Savannah." *American Preservation* 2, 3
     (1979): 9-25.

142. Mattera, P. "Hot Child in the City: Urban Crisis, Urban
     Renaissance and Urban Struggle." *Radical America*
     13, 5 (1979): 49-62.

143. Maurice, A.J. "The Census Undercount: Effects on Federal
     Aid to Cities." *Urban Affairs Quarterly* 17, 3 (1982):
     251-284.

144. Mayo, J.M. "Effects of Sheet Forms on Suburban Neighbor-
     ing Behavior." *Environment and Behavior* 11, 3 (1979):
     375-398.

145. Melosi, M.V. *Pollution and Reform in American Cities*.
     Austin, TX: University of Texas Press, 1980.

146. Miller, R. "Household Activity Patterns in Nineteenth-
     Century Suburbs: A Time-Geographic Exploration."
     Annals of the Association of American Geographers
     72, 3 (1982): 355-370.

147. Mitchelson, R.L. "The Effect of Social Heterogeneity within Urban Travel Corridors on the Travel Behavior of Residents." *Professional Geographer* 34, 2 (1982): 185-196.

148. Molotch, H. "Capital and Neighborhood in the United States: Some Conceptual Links." *Urban Affairs Quarterly* 14, 3 (1979): 289-312.

149. Monkkonen, E.H. "From Cop History to Social History: The Significance of the Police in American History." *Journal of Social History* 15, 4 (1982): 575-592.

150. ————. *Police in Urban America, 1860-1920.* Cambridge, Eng.: Cambridge University Press, 1981.

151. Morris, R.J. *Bum Rap on America's Cities: The Real Causes of Urban Decay.* Englewood Cliffs, NJ: Prentice-Hall, Inc., 1980.

152. Moscato, D.R. "An Empirical Study of Public Sector Employment Patterns of Major Cities of the United States." *Urban Systems* 4, 1 (1979): 27-34.

153. Mowry, G.E., and B.A. Brownell. *The Urban Nation 1920-1980.* New York: Hill and Wang, 1981.

154. Muller, P.O. *Contemporary Suburban America.* Englewood Cliffs, NJ: Prentice-Hall, Inc., 1981.

155. ————. "Everyday Life in Suburbia: A Review of Changing Social and Economic Forces That Shape Daily Rhythms within the Outer City." *American Quarterly* 34, 3 (1982): 262-277.

156. Muller, T., et al. *The Urban Household in the 1980's: A Demographic and Economic Perspective.* Washington, DC: Urban Institute, 1981.

157. Nachmias, C. "Community and Individual Ethnicity: The Structural Context of Economic Performance." *American Journal of Sociology* 85, 3 (1979): 640-652.

158. Nash, G.B. *The Urban Crucible: Social Change, Political Consciousness, and the Origins of the American Revolution.* Cambridge, MA: Harvard University Press, 1979.

159. Newcomer, K., and S. Welch. "The Impact of General Revenue Sharing on Spending in Fifty Cities." *Urban Affairs Quarterly* 18, 1 (1982): 131-144.

160. Norton, R.D. *City Life Cycles and American Urban Policy.* New York: Academic Press, 1979.

161. Oldham, S.G. "Historic Preservation Tax Incentives." *Urban Land* 38, 11 (1979): 3-10.

162. Orski, C.K. "What Can We Learn from Foreign Cities?
      An Effort to Make Cities Livable." *Vital Speeches*
      45, 15 (1979): 469-473.

163. ————. "What Foreign Cities Can Teach American Cities."
      Urban Land 38, 2 (1979): 13-18.

164. Palm, R. *The Geography of American Cities.* New York:
      Oxford University Press, 1981.

165. Patel, D.I. *Exurbs: Urban Residential Developments in
      the Countryside.* Lanham, MD: University Press of
      America, 1980.

166. Peshkin, A. *Growing Up American: Schooling and the
      Survival of Community.* Chicago, IL: University of
      Chicago Press, 1978.

167. Peterson, J.A. "Environment and Technology in the Great
      City Era of American History." *Journal of Urban His-
      tory* 8, 3 (1982): 343-354.

168. ————. "The Impact of Sanitary Reform upon American
      Urban Planning, 1840-1890." *Journal of Social
      History* 13, 1 (1979): 83-104.

169. Platt, D.C. "Financing the Expansion of Cities, 1860-
      1914." *Urban History Review* 11, 3 (1983): 61-67.

170. Podelefsky, A., and F. DuBau. *Strategies for Community
      Crime Prevention: Collective Responses to Crime in
      Urban America.* Springfield, IL: Charles C. Thomas,
      1981.

171. Pollakowski, H.O. *Urban Housing Markets and Residential
      Location.* Lexington, MA: Lexington Books, 1982.

172. Pommer, R. "The Architecture of Urban Housing in the
      United States during the Early 1930's." *Journal of
      the Society of Architectural Historians* 37, 4 (1978):
      235-264.

173. Popenoe, D. "Urban Sprawl: Some Selected Sociological
      Considerations." *Sociology and Social Research* 63, 2
      (1979): 255-268.

174. Powers, R.W. "Cleaning Up Cities." *Current Municipal
      Problems* 5, 3 (1979): 360-364.

175. Pred, A. *Urban Growth and City Systems in the United
      States, 1840-1860.* Cambridge, MA: Harvard University
      Press, 1980.

176. Rafter, O. *The Theory and Practice of Neighborhood
      Planning in the 1970's: A Comprehensive Bibliography.*
      Chicago, IL: Council of Planning Librarians, 1978.

177. Reed, J.S. "Sociology and Regional Studies in the
     United States." *Ethnic and Racial Studies* 3, 1
     (1980): 40-51.

178. Reps, J.W. *Town Planning in Frontier America*. Columbia,
     MO: University of Missouri Press, 1981.

179. Ridgeway, J. *Energy-Efficient Community Planning: A
     Guide to Saving Energy and Producing Power at the
     Local Level*. Emmaus, PA: JG Press, Inc., 1979.

180. Riefter, R.F. "Nineteenth-Century Urbanization Pat-
     terns in the United States." *Journal of Economic
     History* 39, 4 (1979): 951-974.

181. Roesch, M. "Suburban Planning." *Inland Architect* 23, 6
     (1979): 6-11.

182. Roeseler, W.G. *Successful American Urban Plans*. Lex-
     ington, MA: D.C. Heath and Company, 1982.

183. Roneek, D.W., et al. "Female-Headed Families: An Eco-
     logical Model of Residential Concentration in a Small
     City." *Journal of Marriage and the Family* 42, 1
     (1980): 157-170.

184. Rose, J.G. "Exclusionary Zoning in the Federal Courts."
     Zoning and Planning Law Report 2, 7 (1979): 8.

185. Rossell, C. "White Flight: Pros and Cons." *Social
     Policy* 9, 3 (1978): 46-51.

186. Roth, L.M. "Three Industrial Towns by McKim, Mead, and
     White." *Journal of the Society of Architectural His-
     torians* 38, 4 (1979): 317-347.

187. Savas, E.S. "How Much Do Government Services Really
     Cost?" *Urban Affairs Quarterly* 15, 1 (1979): 23-41.

188. Schaffer, D. *Garden Cities for America: The Radburn
     Experience*. Philadelphia, PA: Temple University
     Press, 1983.

189. Schmidt, D.E., et al. "Perceptions of Crowding: Pre-
     dicting at the Residence, Neighborhood, and City
     Levels." *Environment and Behavior* 11, 1 (1979):
     105-130.

190. Schonberg, S.P. "Criteria for the Evaluation of Neigh-
     borhood Vitality in Working-Class and Poor Areas in
     Core Cities." *Social Problems* 27, 1 (1979): 69-78.

191. Schwyler, R.L. *Urban Archaeology in America*. Farming-
     dale, NY: Baywood Publishing Co., 1980.

192. Scobie, I.W. "Family and Community History Through Oral History." *Public Historian* 1, 4 (1979): 29-39.

193. Seiler, L.H., and G.F. Summers. "Corporate Involvement in Community Affairs." *Sociological Quarterly* 20, 3 (1979): 375-386.

194. Shakow, D. "The Municipal Farmers Market as an Urban Service." *Economic Geography* 57, 1 (1981): 68-77.

195. Shonick, W. "The Public Hospital and Its Local Ecology in the U.S. Some Relationships Between the Plight of the Public Hospital and the Plight of the Cities." *International Journal of Health Services* 9, 3 (1979): 359-396.

196. Siegch, A. *The Image of the American City in Popular Literature, 1820-1870.* Port Washington, NY: Kennikat Press, 1981.

197. Smith, H.H. *The Citizen's Guide to Planning.* Chicago, IL: American Planning Association, 1979.

198. Entry deleted.

199. Solomon, A.P. *The Prospective City.* Cambridge, MA: MIT Press, 1980.

200. South, S.J., and D.L. Poston. "A Note on Stability in the U.S. Metropolitan System, 1950-1970." *Demography* 17, 4 (1980): 445-450.

201. Squires, G.D., et al. "Urban Decline or Disinvestment: Uneven Development, Redlining and the Role of the Insurance Industry." *Social Problems* 27, 1 (1979): 79-95.

202. Stamm, C.F. "Urban Fiscal Stress: Is It Inevitable?" *Tempo* 25, 1 (1979): 13-18.

203. Steinnes, D.N. "Suburbanization and the Malling of America: A Time-Series Approach." *Urban Affairs Quarterly* 17, 4 (1982): 401-418.

204. Sternlieb, G., and D. Listokin. *New Tools for Economic Development: The Enterprise Zone, Development Bank and RFC.* New Brunswick, NJ: Rutgers University Press, 1981.

205. Stillman, R.J. *The Rise of the City Manager: A Public Professional in Local Government.* Albuquerque, NM: University of New Mexico Press, 1979.

206. Stimpson, R., et al. *Women and the American City.* Chicago, IL: University of Chicago Press, 1980.

207. Strom, F.A. "Legal Standing of Citizen Groups." *Zoning and Planning Law Report* 2, 11 (1979): 169-174.

208. Struyk, R.J. *A New System for Public Housing: Salvaging a National Resource*. Washington, DC: The Urban Institute, 1980.

209. ———, and B.J. Soldo. *Improving the Elderly's Housing: A Key to Preserving the Nation's Housing Stock and Neighborhoods*. Cambridge, MA: Ballinger Publishing Co., 1979.

210. Susser, I. *Norman Street: Poverty and Politics in Our Urban Neighborhood*. New York: Oxford University Press, 1982.

211. Swenson, B.E., et al. *Small Towns and Small Towners: A Framework for Survival and Growth*. Beverly Hills, CA: Sage Publications, 1979.

212. Symes, M. "Urban Design Education in Britain and America." *Education for Urban Design*. Purchase, NY: Institute for Urban Design, 1982.

213. Taylor, D.W. "The Assessment of Public Agency Responsiveness: A Pilot Study in an Urban Context." *Urban Systems* 4, 3/4 (1979): 243-254.

214. Thomas, R.D. "Implementing Federal Programs at the Local Level." *Political Science Quarterly* 94, 3 (1979): 419-436.

215. Tise, L.E. "State and Local History: A Future from the Past." *Public Historian* 1, 4 (1979): 14-22.

216. "Toward an Interdisciplinary History of the American City." *Journal of Urban History* 8, 4 (1982): 447-485.

217. Turner, L.T., et al. *Review of Policies Affecting the Housing Decisions of Older Americans*. Iowa City, IA: University of Iowa Press, 1981.

218. Vajs, K. *Redevelopment in Minneapolis, 1960-1977*. Chicago, IL: Council of Planning Librarians, 1977.

219. Varaiya, P., and M. Wiseman. "Investment and Employment in Manufacturing in United States Metropolitan Areas, 1960-1976." *Regional Science and Urban Economics* 11, 4 (1981): 431-470.

220. Waddell, J.O., and O.M. Watson. *The American Indian in Urban Society*. Lanham, MD: University Press of America, 1984.

221. Wade, R. "America's Cities Are (Mostly) Better Than
     Ever." *American Heritage* 30, 2 (1979): 4-13.

222. Weaver, C., and R.F. Babcock. "City Zoning: The Once
     and Future Frontier." *Planning* 45, 12 (1979): 19-23.

223. Weinberg, N. *Preservation in American Cities and
     Towns*. Boulder, CO: Westview Press, 1979.

224. Whitehead, J.W.R. *The Urban Landscape: Historical
     Development and Management*. New York: Academic
     Press, 1982.

225. Whyte, W.F. *Street Corner Society*. 3rd ed. Chicago, IL:
     University of Chicago Press, 1981.

226. Wilson, M.G. *The American Woman in Transition: The
     Urban Influence: 1870-1920*. Westport, CT: Greenwood
     Press, 1979.

227. Wilson, W.H. "Horace McFarland and the City Beautiful
     Movement." *Journal of Urban History* 7, 3 (1981):
     315-334.

228. Wolfe, M.R. *Lucius Polk Brown and Progressive Food
     and Drug Control: Tennessee and New York City 1908-
     1920*. Lawrence, KS: Regents Press of Kansas, 1978.

229. Yeates, M. *North American Urban Patterns*. New York:
     Halstead Press, 1980.

230. Yin, R.K. *Conserving America's Neighborhoods*. New
     York: Plenum Press, 1982.

THE EAST

General

231. Carlino, G.A. "Contrasts in Agglomeration: New York
and Pittsburgh Reconsidered." *Urban Studies* 17, 3
(1981): 343-352.

232. Cheape, C.W. *Moving the Masses: Urban Public Transit
in New York, Boston, and Philadelphia, 1880-1912.*
Cambridge, MA: Harvard University Press, 1980.

233. Condran, G., and E. Crimmins. "Mortality Differentials
Between Rural and Urban Areas of States in the North-
eastern United States." *Journal of Historical Geography* 6, 2 (1980): 179-202.

234. Conzen, M.P. "American Cities in Profound Transition:
The New City Geography of the 1980's." *Journal of
Geography* 82, 3 (1983): 94-102.

235. Foley, J.W. "Community Structure and Public Policy
Outputs in 300 Eastern American Communities: Toward
a Sociology of the Public Sector." *Ethnicity* 6, 3
(1979): 222-234.

236. Greer, W. "Employment Growth in New York and New Jersey: The Effects of Suburbanization." *Federal Re-
serve Bank of New York* 7, 3 (1982): 48-53.

237. Liroff, R.A., and G.G. Davis. *Protecting Open Space:
Land Use Control in the Adirondack Park.* Cambridge,
MA: Ballinger Publishing Co., 1981.

238. Maier, P. *Boston and New York in the Eighteenth Century.* Charlottesville, VA: University Press of Virginia, 1981.

239. Miller, M.L., and J. van Maanen. "Getting into Fishing:
Observations on the Social Identities of New England
Fishermen." *Urban Life* 11 (April 1982): 27-54.

240. Wood, J.S. "Village and Community in Early Colonial
New England." *Journal of Historical Geography* 8, 4
(1982): 333-346.

New England

241. Alexopoulas, J. "The Creator of Bushell Park." *Con-
     necticut Historical Society Bulletin* 47, 3 (1982):
     65-73.

242. Bartlett, I.H. "Abolitionists, Fugitives, and Imposters
     in Boston, 1846-1847." *New England Quarterly* 55, 1
     (1982): 97-110.

243. Blodgett, G. "Yankee Leadership in a Divided City:
     Boston 1860-1910." *Journal of Urban History* 8, 4
     (1982): 371-389.

244. Bucki, C.F. "Dilution and Craft Tradition: Bridgeport,
     Connecticut, Munitions Workers, 1915-1919." *Social
     Science History* 4, 1 (1980): 105-124.

245. Buell, E.H. "Busing and the Defended Neighborhood:
     South Boston, 1974-1977." *Urban Affairs Quarterly*
     16, 2 (1980): 161-188.

246. Cooke, E.S. "The Boston Furniture Industry in 1880."
     Old-Time New England 70, 257 (1980): 82-98.

247. Cumbler, J.T. *A Moral Response to Industrialism: The
     Lectures of Reverend Cook in Lynn, Massachusetts.*
     Albany, NY: State University of New York Press, 1981.

248. Early, F.H. "The French-Canadian Family Economy and
     Standard-of-Living in Lowell, Massachusetts, 1870."
     Journal of Family History 7, 2 (1982): 180-199.

249. Faler, P.G. *Mechanics and Manufacturers in the Early
     Industrial Revolution: Lynn, Massachusetts, 1780-1860.*
     Albany, NY: State University of New York Press, 1981.

250. Ferber, M., and E. Bead. "Marketing Urban America: The
     Selling of the Boston Plan and a New Direction in
     Federal Urban Relations." *Polity* 12, 4 (1980): 539-
     559.

251. Garland, J.E. *Boston's Gold Coast: The North Shore,
     1890-1929.* Boston, MA: Little, Brown and Company,
     1982.

252. Gersuny, C. "Industrial Casualties in Lowell, 1890-
     1905." *Labor History* 20, 3 (1979): 435-442.

253. Hardy, S. *How Boston Played: Sport, Recreation and
     Community 1865-1915.* Boston, MA: Northeastern Uni-
     versity Press, 1981.

254. Hartman, C., et al. "Dilemmas of Community Organizing:
     Mission Hill in Boston." *Social Forces* 9, 1 (1978):
     41-52.

255. Hemphill, C.D. "Women in Court: Sex-Role Differentiation in Salem, Massachusetts 1636 to 1683." *William and Mary Quarterly* 39, 1 (1982): 164-175.

256. Jones, D.L. *Village and Seaport: Migration and Society in Eighteenth-Century Massachusetts*. Hanover, NH: New England University Press, 1981.

257. Kay, J.H. *Lost Boston*. Boston, MA: Houghton Mifflin, 1980.

258. Krefetz, S.P. "Low and Moderate-Income Housing in the Suburbs: The Massachusetts' Anti-Snob Zoning Law." *Policy Studies Journal* 8, 2 (1979): 288-299.

259. Lenger, F. "Class, Culture and Class Consciousness in Antebellum Lynn: A Critique of Alan Dawley and Paul Faler." *Social History* 6, 3 (1981): 317-332.

260. Lovett, R.W. "Remembered Triangle or Growing up in the North Beverly in the 1920's." *Essex Institute Historical Collections* 117, 4 (1981): 283-295.

261. Mudgett, D. "Bringing Excitement Back to Baystate West." *Urban Land* 42, 1 (1983): 22-25.

262. Nash, G.B. "The Failure of Female Factory Labor in Colonial Boston." *Labor History* 20, 2 (1979): 165-188.

263. Page, J. "The Economic Structure of Society in Revolutionary Bennington." *Vermont History* 49, 2 (1981): 69-84.

264. Pearson, R.L., and L. Wrigley. "Before Mayor Richard Lee: George Dudley Seymour and the City Planning Movement in New Haven, 1907-1924." *Journal of Urban History* 6, 3 (1980): 297-319.

265. Pease, J.H., and W.H. Pease. "Social Structure and the Potential for Urban Change: Boston and Charleston in the 1830's." *Journal of Urban History* 8 (February 1982): 171-196.

266. ———. "Paternal Dilemmas: Education, Property, and Patrician Persistence in Jacksonian Boston." *New England Quarterly* 53, 2 (1980): 147-167.

267. Pencak, W. "The Social Structure of Revolutionary Boston: Evidence from the Great Fire of 1760." *Journal of Interdisciplinary History* 10, 2 (1979): 267-279.

268. Ryan, R.J. "Boston Rediscovered." *Challenge* 10, 9 (1979): 16-25.

269. Schlichteng, K. "Decentralization and the Decline of
      the Central City: The Case of Bridgeport." *American
      Journal of Economics and Sociology* 40, 4 (1981): 353–
      366.

270. Silverman, R.A. *Law and Urban Growth, Civil Litigation
      in the Boston Trial Courts, 1800-1900.* Princeton, NJ:
      Princeton University Press, 1981.

271. Simmons, J.A. "Land Bank Deals: Massachusetts New De-
      velopments." *Urban Land* 41, 7 (1982): 10–17.

272. Stave, S.A. *Hartford, the City and the Region: Past,
      Present, Future.* Hartford, CT: University of Hartford
      Press, 1979.

273. Steen, I.D. "Cleansing the Puritan City: The Reverend
      Henry Morgan's Antivice Crusade in Boston." *New
      England Quarterly* 54, 3 (1981): 385–441.

274. Story, R. *The Forging of an Aristocracy: Harvard and
      the Boston Upperclass, 1800-1870.* Middletown, CT:
      Wesleyan University Press, 1980.

275. Taggart, H., et al. "Redlining: An Assessment of the
      Evidence of Disinvestment in Metropolitan Boston."
      Urban Affairs 17, 1 (1981): 91–114.

276. Ueda, R. "Suburban Social Change and Educational Re-
      form: The Case of Somerville, Massachusetts, 1912-
      1924." *Social Science History* 3, 3/4 (1979): 167–203.

277. Useem, B. "Models of the Boston Anti-Busing Movement:
      Polity/Mobilization and Relative Deprivation." *Socio-
      logical Quarterly* 22, 2 (1981): 263–274.

278. ———. "Trust in Government and the Boston Anti-
      Busing Movement." *Western Political Quarterly* 35, 1
      (1982): 81–91.

Middle Atlantic States

279. Aimone, A.C., and E. Manders. "A Note of New York's
      Independent Companies." *New York History* 63 (January
      1982): 59–73.

280. Alberts, R.C. *The Shaping of the Point: Pittsburgh's
      Renaissance Park.* Pittsburgh, PA: University of Pitts-
      burgh Press, 1981.

281. Arnold, J.L. "The Neighborhood and the City Hall: The
      Origin of Neighborhood Associations in Baltimore,
      1880-1911." *Journal of Urban History* 6, 1 (1979): 3–30.

282. Baritz, L. "The Culture of Manhattan." *The Massachusetts Review* 22 (Autumn 1981): 403-422.

283. Bauman, J.F. "Downtown Versus Neighborhood: Focusing on Philadelphia in the Metropolitan Era, 1920-1980." *Pennsylvania History* 48, 1 (1980): 3-20.

284. ————. "Visions of a Postwar City: A Perspective on Urban Planning in Philadelphia and the Nation, 1942-1945." *Urbanism Past and Present* 6, 1 (1981): 1-11.

285. Baumann, R. "Philadelphia's Manufacturers and the Excise Taxes of 1794: The Forging of the Jeffersonian Coalition." *Pennsylvania Magazine of History and Biography* 106, 1 (1982): 3-39.

286. Becker, L.L. "The People and the System in a Colonial Pennsylvania Town." *Pennsylvania Magazine* 105, 2 (1981): 135-149.

287. Beirne, D.R. "Hampden-Woodberry: The Mill Village in an Urban Setting." *Maryland Historical Magazine* 77, 1 (1982): 6-26.

288. ————. "Late Nineteenth-Century Industrial Communities in Baltimore." *Maryland Historian* 11, 1 (1980): 39-50.

289. Borchert, J. "Alley Landscapes of Washington." *Landscape* 23, 3 (1979): 3-10.

290. ————. *Alley Life in Washington: Family Community and Folklife in the City, 1850-1970.* Chicago, IL: University of Illinois Press, 1980.

291. Brandow, J.C. "Notes and Documents: A Barbados Planter's Visit to Philadelphia in 1837: The Journal of Nathaniel T.W. Carrington." *Pennsylvania Magazine of History and Biography* 106, 3 (1982): 411-422.

292. Browne, G.L. *Baltimore in the Nation 1789-1861.* Chapel Hill, NC: University of North Carolina Press, 1980.

293. Carey, G.W. "New York: World Economy, Federal Politics." *Focus* 31, 4 (1981): 1-16.

294. Cashdan, L., et al. "Roses from Rubble: New Uses for Vacant Urban Land." *New York Affairs* 7, 2 (1982): 89-96.

295. Capeci, D.J. "Fiorello H. La Guardia and the American Dream: A Document." *Italian Americana* 4, 1 (1978): 1-21.

296. ————. "Fiorello H. La Guardia and the Stuyvesant Town

Controversy of 1943." *The New-York Historical Society Quarterly* 42, 4 (1978): 289-310.

297. Chang, J.M. "No Refuge in the Bronx." *City Limits* 7, 10 (1982): 12-17.

298. "The City's Economic Development Plan." *Neighborhoods* (Philadelphia) 5, 1, 17-27.

299. Clement, P.F. "The Philadelphia Welfare Crisis of the 1820's." *Pennsylvania Magazine* 105, 2 (1981): 150-165.

300. Cochran, T.C. "Philadelphia: The American Industrial Center, 1750-1850." *Pennsylvania Magazine of History and Biography* 106, 3 (1982): 323-340.

301. Condit, C.W. *The Port of New York: A History of the Rail and Terminal System from the Beginnings to Pennsylvania Station.* Chicago, IL: University of Chicago Press, 1980.

302. Condran, G.A., and R.A. Cheny. "Mortality Trends in Philadelphia: Age and Case Specific Death Rates 1870-1930." *Demography* 19, 1 (1982): 97-124.

303. Connolly, H.X. *A Ghetto Grows in Brooklyn.* New York: New York University Press, 1977.

304. Cox, R.J. "Understanding the Monumental City: A Bibliographical Essay on Baltimore History." *Maryland History Magazine* 77, 1 (1982): 70-111.

305. Cutler, W.W., and H. Gillette. *The Divided Metropolis: Social and Spatial Dimensions of Philadelphia, 1800-1975.* Westport, CT: Greenwood Press, 1980.

306. Damm, D., et al. "Response of Urban Real Estate Values in Anticipation of the Washington Metro." *Journal of Transport and Economics and Policy* 14, 3 (1980): 315-326.

307. Daniels, T. "A Philadelphia Squatters' Shack: Urban Pioneering." *Pioneer America* 13, 2 (1981): 43-48.

308. Davies, E.J. "Large-Scale Systems and Regional Leadership: Wilkes-Barre's Upper Class and Urban Elites in the Northern Anthracite Region, 1920-1930." *Public Historian* 4, 4 (1982): 39-69.

309. Davis, S.G. "Making Night Hideous: Christmas Revelry and Public Disorder in Nineteenth-Century Philadelphia." *American Quarterly* 34 (Summer 1982): 185-199.

310. Downs, A. "The Necessity for Neighborhood Deterioration." *New York Affairs* 7, 2 (1982): 35-40.

311. Dye, N.S. *As Equals and as Sisters: Feminism, the Labor Movement, and the Women's Trade Union League of New York.* Columbia, MO: University of Missouri Press, 1980.

312. Ellbott, P., and W. Kempey. "New York City's Tax Abatement and Exemption Program for Encouraging Housing Rehabilitation." *Public Policy* 26, 4 (1978): 571-598.

313. Erickson, R.A. "Employment Density Variation in the Baltimore Metropolitan Area." *Environment and Planning A* 14, 5 (1982): 591-602.

314. Evans, P.F. *City Life--A Perspective from Baltimore, 1968-1978.* Columbia, MD: Fairfax, 1981.

315. Fickett, L. "Wooddale: An Industrial Community." *Delaware History* 19, 4 (1981): 229-242.

316. Fishbein, L. "The Failure of Feminism in Greenwich Village before World War I." *Women's Studies* 9, 3 (1982): 275-290.

317. Folsom, B.W. *Urban Capitalists, Entrepreneurs, and City Growth in Pennsylvania's Lackawanna and Lehigh Regions, 1800-1920.* Baltimore, MD: Johns Hopkins University Press, 1981.

318. Froncek, T. *The City of Washington: An Illustrated History.* New York: Alfred A. Knopf, 1977.

319. Glazer, N. "The Urban Dilemma: New York." *Ekistics* 46, 275 (1979): 72-75.

320. Goldfield, D.R. "Private Neighborhood Redevelopment and Displacement: The Case of Washington D.C." *Urban Affairs Quarterly* 15, 4 (1980): 453-468.

321. Gonzalez, N.L. "Garifuna Settlement in New York: A New Frontier." *International Migration Review* 13, 2 (1979): 255-263.

322. Greene, S.E. *Baltimore: An Illustrated History.* Woodland Hills, CA: Windsor Publications, 1980.

323. Gregory, P.R. *Parks and People, Values, and Decisions: Proposals for the Breezy Point Park, New York City.* Ithaca, NY: Cornell University Press, 1980.

324. Grinder, R.D. "From Insurgency to Efficiency: The Smoke Abatement Campaign in Pittsburgh before World War I." *The Western Pennsylvania Historical Magazine* 61, 3 (1978): 187-202.

325. Guida, L., and S. Oristaglio. "Powelton Village: One of

Philadelphia's Oldest Neighborhoods Experiences a
Rebirth." *American Preservation* 2, 2 (1979): 44-52.

326.   Hall, B.W.   "Elites and Spatial Change in Pittsburgh:
       Minersville as a Case Study." *Pennsylvania History*
       48, 4 (1981): 311-334.

327.   Hammack, D.C.   *Power and Society: Greater New York at
       the Turn of the Century*. New York: Basic Books,
       1982.

328.   Heavner, R.O.   "Indentured Servitude: The Philadelphia
       Market, 1771-1773." *The Journal of Economic History*
       38, 3 (1978): 701-713.

329.   Hershberg, T.   *Philadelphia: Work, Space, Family, and
       Group Experience in the Nineteenth Century*. New York:
       Oxford University Press, 1981.

330.   Hosay, P.M.   *The Challenge of Urban Poverty: Charity
       Reformers in New York City, 1835-1890*. New York:
       Arno Press, 1980.

331.   Houstan, D.   "A History of the Process of Capital Accu-
       mulation in Pittsburgh: A Marxist Interrogation, Pt.
       3." *Review of Regional Studies* 10, 2 (1980): 20-37.

332.   Hudson, R.   "Accumulation, Spatial Policies, and the
       Production of Regional Labor Reserves: A Study of
       Washington New Town." *Environment and Planning A*
       14, 5 (1982): 665-680.

333.   Hurst, H.W.   "The Northernmost Southern Town: A Sketch
       of Pre-Civil War Annapolis." *Maryland Historical
       Magazine* 76, 3 (1981), 240-249.

334.   Hurst, M., and R.E. Zambrana.   "The Health Careers of
       Urban Women: A Study in East Harlem." *Signs* 5, 3
       (1980): 117-126.

335.   Hyser, R.M.   "Discord in Utopia: The Ellsworth Strike
       of 1904." *Pennsylvania Magazine of History and Biog-
       raphy* 106, 3 (1982): 393-410.

336.   Jentz, J.B.   "The Anti-Slavery Constituency in Jack-
       sonian New York City." *Civil War History* 27, 2
       (1981): 101-122.

337.   Johnson, P.E.   *A Shopkeeper's Millennium: Society and
       Revivals in Rochester, New York 1815-1837*. New York:
       Hill and Wang, 1978.

338.   Jucha, R.J.   "The Anatomy of a Streetcar Suburb: A
       Development History of Shadyside, 1852-1916." *Western
       Pennsylvania Historical Magazine* 62, 4 (1979): 301-319.

339. Kantrow, L. "Philadelphia Gentry: Fertility and Family Limitation among an American Aristocracy." *Population Studies: A Journal of Demography* 34, 1 (1980): 21-30.

340. Kellner, G.H., and J.S. Lemons. "Providence: A Century of Greatness." *Rhode Island History* 41, 1 (1982): 3-18.

341. Kessner, T. "Jobs, Ghettoes and the Urban Economy, 1880-1935." *American Jewish History* (New York) 71, 2 (1981): 218-238.

342. Ketchum, C.G. "Some Interesting Pittsburghers, 1911-1941." *Western Pennsylvania Historical Magazine* 65, 2 (1982): 99-115.

343. Kim, S.B. "Impact of Class Relations and Warfare in the American Revolution: The New York Experience." *Journal of American History* 69, 2 (1982): 326-346.

344. Kinsey, D.N. "A Partnership for Casino Development: State Environmental Perspectives on Atlantic City's Renaissance." *Environmental Comment* (August 1979): 4-13.

345. Klebaner, B.J. *New York City's Changing Economic Base.* New York: Pica Press, 1981.

346. Klem, M.M. "An Experiment That Failed: General James Robertson and Civil Government in British New York, 1779-1783." *New York History* 61, 3 (1980): 229-254.

347. Kraus, H.P. *The Settlement House Movement in New York City, 1886-1914.* New York: Arno Press, 1980.

348. Kross, J. *The Evolution of an American Town: New Town, New York 1642-1775.* Philadelphia, PA: Temple University Press, 1983.

349. Lacey, B.E. "Women in the Era of the American Revolution: The Case of Norwich, Connecticut." *New England Quarterly* 53, 4 (1980): 527-543.

350. Landsberg, H.E. "Atmospheric Changes in a Growing Community (Columbia, Maryland)." *Urban Ecology* 4, 1 (1979): 53-82.

351. Lane, R. *Violent Death in the City: Suicide, Accident, and Murder in Nineteenth-Century Philadelphia.* Cambridge, MA: Harvard University Press, 1979.

352. Laurie, B. *Working People in Philadelphia, 1800-1850.* Philadelphia, PA: Temple University Press, 1980.

353. Lawson, R. "Origins and Evolution of a Social Movement

Strategy: The Rent Strike in New York City: 1904–
1980." *Urban Affairs Quarterly* 18, 3 (1983): 371–
396.

354. Lightfoot, F.S. *Nineteenth-Century New York in Rare
Photographic Views*. New York: Constable, 1981.

355. Llewellyn, R. *Washington: The Capital*. Washington,
DC: The Preservation Press, 1981.

356. Lukacs, J.A. *Philadelphia, Patricians and Philistines,
1900-1950*. New York: Farrar, Straus and Giroux, 1981.

357. Lyons, P. *Philadelphia Communists, 1936-1956*. Philadelphia, PA: Temple University Press, 1982.

358. Mandelbaum, S. *Boss Tweed's New York*. Westport, CT:
Greenwood Press, 1982.

359. Mannix, J.M. "Philadelphia Constructs Commuter Rail
Connection." *National Railway Bulletin* 46, 6 (1981):
20-23; 46.

360. Marks, B.E. "Rural Response to Urban Penetration: Baltimore and St. Mary's, Maryland, 1790-1840." *Journal
of Historical Geography* 8 (April 1982): 113-128.

361. Miller, C.G. "Local History as a Guide to Planning: A
Case Study of Trumansburg, New York." *Small Town*
10, 1-6 (1979): 4-12.

362. Miller, E.W. "Pittsburgh: Patterns of Evolution."
Pennsylvania Geographer 29, 3 (1981): 6-19.

363. Miller, F. "Documenting Modern Cities: The Philadelphia Model." *Public Historian* 5, 2 (1983): 75-86.

364. Miller, R.B. *City and Hinterland: A Case Study of Urban
Growth and Regional Development* [Syracuse, NY]. Westport, CT: Greenwood Press, 1979.

365. Miller, R.S. *Brooklyn USA*. New York: Columbia University Press, 1979.

366. Morris, C.R. *The Cost of Good Intentions: New York
City and the Liberal Experiment*. New York: W.W.
Norton Co., 1980.

367. Mushkat, J. *The Reconstruction of the New York Democracy*. Madison, NJ: Fairleigh Dickinson University
Press, 1981.

368. Neches, A.J., and P. Aarons. "The City Approaches Industrial Development." *New York Affairs* 6, 2 (1980):
43-46.

369. Olsen, S. "Yankee City and the New Urban History." *Journal of Urban History* 6, 3 (1980): 321-337.

370. Entry deleted.

371. Olson, S.H. "Baltimore Imitates the Spider." *Annals of the Association of American Geographers* 69, 4 (1979): 557-574.

372. ————. *Baltimore: The Building of an American City.* Baltimore, MD: Johns Hopkins University Press, 1980.

373. Olstein, A. "Park Slope: The Warren Street Balancing Act." *New York Affairs* 7, 2 (1982): 59-65.

374. Paris, A. "Hidden Dimensions of the New York City Fiscal Crisis." *Review of Black Political Economy* 10, 3 (1980): 262-278.

375. Patrick, L.L. "A Geography of Recline: The Bedding Industry of Pittsburgh, Pennsylvania." *Pennsylvania Geographer* 19, 3 (1981): 20-34.

376. Pierce, N., and J. Hagstrom. "Renewed, 1979 Style, in the South Bronx." *National Journal* 40 (October 6, 1979): 1644-1648.

377. Pellicano, R.R. "Teacher Unionism in New York City." *Urban Education* 17 (April 1982): 97-116.

378. Perlman, J. "New York from the Bottom Up." *New York Affairs* 7, 2 (1982): 27-35.

379. Perry, E.I. "Industrial Reform in New York City: Belle Moskowitz and the Protocol of Peach, 1913-1916." *Labor History* 23 (Winter 1982): 5-31.

380. Pessen, E. "Those Marvelous Depression Years Reminiscences of the Big Apple." *New York History* 62, 2 (1981): 188-200.

381. Preston, R.M. "The Great Fire of Emmitsburg, Maryland: Does a Catastrophic Event Cause Mobility?" *Maryland Historical Magazine* 77, 2 (1982): 172-182.

382. Rondinelli, D.A. "An Assessment of the Neighborhood Housing Services Model for Neighborhood Reinvestment in Syracuse, New York State." *Ekistics* 50, 298 (1983): 14-24.

383. Ryan, M.P. *Cradle of the Middle Class: The Family in Oneida County, New York, 1790-1865.* Cambridge, Eng.: Cambridge University Press, 1981.

384. Sanders, R.A. "Diversity in the Street Trees of Syracuse, New York." *Urban Ecology* 5, 1 (1981): 33-44.

385. Schreiber, L. "Bluebloods and Local Societies: A Philadelphia Microcosm." *Pennsylvania History* 48, 3 (1981): 251-266.

386. Schwartz, S. "Cantwell's Bridge, Delaware: A Demographic and Community Study." *Delaware History* 19, 1 (1980): 20-38.

387. Seey, B.F. "Wilmington and Its Railroads: A Lasting Connection." *Delaware History* 19, 1 (1980): 1-19.

388. Shirely, R.I. "Pittsburgh: Renaissance, Redevelopment and Revitalization, 1945-1981." *Pennsylvania Geographer* 19, 3 (1981): 35-45.

389. Simmons, D.L., and R.J. Reynolds. "Effects of Urbanization on Base Flow of Selected South-Shore Streams, Long Island, New York." *Water Resources Bulletin* 18, 5 (1982): 797-807.

390. Simpson, M. "Meliorist versus Insurgent Planners and the Problems of New York, 1921-1941." *Journal of American Studies* 16, 2 (1982): 207-228.

391. Singh, U.P., et al. "Computer Optimized Stormwater Treatment (Cost) Program: Philadelphia Case Program." *Water Resources Bulletin* 18, 5 (1982): 769-778.

392. Slater, M. "Flatbush: Citibank's Test Flight." *New York Affairs* 7, 12 (1982): 40-50.

393. Spann, E.K. *The New Metropolis: New York City, 1840-1857.* New York: Columbia University Press, 1981.

394. Stansell, C. "Women, Children, and the Uses of the Streets: Class and Gender Conflict in New York City, 1850-1860." *Feminist Studies* 8, 2 (1982): 309-336.

395. Stegman, M.A. *The Dynamics of Rental Housing in New York City.* Piscataway, NJ: Center for Urban Policy Research, 1982.

396. Steiner, F. "Philadelphia: The Holy Experiment." *Ekistics* 49, 295 (1982): 298-305.

397. Steller, J.D. "A MD Takes Off: Baltimore's Inner Harbor." *Urban Land* 41 (March 1981): 1-20.

398. Tabb, W.K. *The Long Default: New York City and the Urban Fiscal Crisis.* New York: Monthly Review Press, 1981.

399. Tauranac, J. *Essential New York: A Guide to the History and Architecture of Manhattan's Important Buildings, Parks, and Bridges.* New York: Holt, Rinehart and Winston, 1979.

400. Thomas, M.W. "Walt Whitman and Manhattan-New York." *American Quarterly* 34, 4 (1982): 362-378.

401. Vill, M.J. "Residential Development on a Landed Estate: The Case of Baltimore's Harlem." *Maryland Historical Magazine* 77, 3 (1982): 266-278.

402. Vitullo-Martin, J. "Why New York City Can't Help Itself." *New York Affairs* 6, 2 (1980): 37-42.

403. Wallace, R. "Contagion and Incubation in New York City Structural Fires, 1964-1976." *Human Ecology* 6, 4 (1978): 423-434.

404. Wallace, R. "Fire Service Productivity and the New York Fire Crises, 1968-1979." *Human Ecology* 9 (December 1981): 433-464.

405. Weinbaum, P.O. *Mobs and Demagogues: The New York Response to Collective Violence in the Early Nineteenth Century.* Ann Arbor, MI: UMI Research Press, 1978.

406. Weslager, C.A. "The City of Amsterdam's Colony on the Delaware, 1656-1664: With Unpublished Dutch Notarial Abstract." *Delaware History* 20, 1 (1983): 1-26.

407. White, D.W. *A Village at War: Chatham, New Jersey and the American Revolution.* Madison, NJ: Fairleigh Dickinson University Press, 1979.

408. White, P.L. *Beekmantown, New York: Forest Frontier to Farm Community.* Austin, TX: University of Texas Press, 1979.

409. Wilentz, S. "Crime, Poverty and the Streets of New York City: The Diary of William H. Bell, 1850-1851." *History Workshop* 7 (Speing 1979): 126-155.

410. Wohlenberg, E.H. "The Geography of Civility Revisited: New York Blackout Looting, 1977." *Economic Geography* 58 (January 1982): 29-44.

411. Wolf, R. *New York.* Washington, DC: The Preservation Press, 1981.

412. Yellowitz, I. *Essays in the History of New York City: A Memorial to Sidney Pomerantz.* Port Washington, NY: Kennikat Press, 1978.

413. Zlotnick, J. *Portrait of an American City: The Novelists' New York.* Port Washington, NY: Kennikat Press, 1982.

414. Zuken, S. "Loft Living as Historic Compromises in the
      Urban Core: The New York Experience." *International
      Journal of Urban and Regional Research* 6, 2 (1982):
      256-267.

THE SOUTH

415. Abbot, C. *The New Urban America: Growth and Politics in the Sunbelt Cities*. Chapel Hill, NC: University of North Carolina Press, 1981.

416. Adams, J.E. "The Welfare Efficiency of Moving Families into Public Housing in Little Rock, Arkansas." *Land Economics* 58 (May 1982): 217-224.

417. Amos, H.E. "City Belles: Images and Realities of the Lives of White Women in Antebellum Mobile." *Alabama Review* 37, 1 (1981): 3-19.

418. Basile, L. "Life in Kosciusko during the 1850's." *Journal of Mississippi History* 42, 4 (1980): 336-346.

419. Beeler, D. "Race Riot in Columbia, Tennessee: February 25-27, 1946." *Tennessee Historical Quarterly* 39, 1 (1980): 49-61.

420. Blackburn, G., and S.L. Richards. "The Mother-Headed Family among Free Negroes in Charleston, South Carolina, 1850-1860." *Phylon* 42, 1 (1981): 11-25.

421. Blasi, Anthony J. *Segregationist Violence and Civil Rights Movement in Tuscoloosa*. Washington, DC: University Press of America, 1980.

422. Bopp, W.J., and M. Wiatrowski. "Police Strike in New Orleans: A City Abandoned by Its Police." *Police Journal* 55, 2 (1982): 125-136.

423. Bullard, R.D., and D.L. Tryman. "Competition for Decent Housing: A Focus on Housing Discrimination Complaints in a Sunbelt City." *Journal of Ethnic Studies* 7, 4 (1980): 51-63.

424. Burman, S. "The Illusion of Progress: Race and Politics in Atlanta, Georgia." *Ethnic and Racial Studies* 2, 4 (1979): 441-454.

425. Cable, M. *Lost New Orleans*. Boston, MA: Houghton Mifflin, 1980.

426. Carvalho, J. "The Baughman Boycott and Its Effect on
     the Richmond, Virginia Labor Movement, 1886-1888."
     Histoire Sociale 12, 24 (1979): 409-417.

427. Catlin, R.A. "Analysis of the Community Development
     Block Grant Program in Nine Florida Cities, 1975-
     1979." *Urban and Social Change Review* 14 (Winter
     1981): 3-11.

428. Chafe, W.H. *Civilities and Civil Rights: Greensboro,
     North Carolina and the Black Struggle for Freedom.*
     New York: Oxford University Press, 1980.

429. Chambers, S.A. *Lynchburg: An Architectural History.*
     Charlottesville, VA: University Press of Virginia,
     1982.

430. Chesson, M.B. *Richmond After the War.* Richmond, VA:
     Virginia State Library, 1981.

431. Cobb, J.C. "Polarization in a Southern City: The
     Augusta Riot and the Emerging Character of the 1970's."
     Southern Studies 20, 2 (1981): 185-200.

432. Coccossis, H.N. *The Shape and Structure of Metropolitan
     Areas: An Application of Gentrographic Analysis to
     Atlanta, Georgia, 1940-1975.* Ithaca, NY: Cornell
     University Press, 1980.

433. Collins, T.W. *Cities in a Larger Context: Southern
     Anthropological Society Proceedings.* Athens, GA:
     University of Georgia Press, 1981.

434. Crump, N.C. "Hopewell during World War I: The Toughest
     Town North of Hell." *Virginia Cavalcade* 31, 1 (1981):
     38-47.

435. Deagan, K. "Downtown Survey: The Discovery of Sixteenth-
     Century St. Augustine in an Urban Area." *American
     Antiquity* 48, 3 (1981): 626-633.

436. ———. "Spanish St. Augustine: America's First Melting
     Pot." *Archaeology* 33, 5 (1980): 22-30.

437. Demetz, K. "Minstrel Dancing in New Orleans' Nineteenth-
     Century Theaters." *Southern Quarterly* 20, 2 (1982):
     28-40.

438. Doyle, D.H. "The Urbanization of Dixie." *Journal of
     Urban History* 7, 1 (1980): 83-91.

439. Drago, E., and R. Melnick. "The Old Slave Mart Museum;
     Charleston, South Carolina: Rediscovering the Past."
     Civil War History 27, 2 (1981): 138-154.

440. Elgie, R.A., and A.R. Clark. "Social Class Segregation in Southern Metropolitan Areas." *Urban Affairs Quarterly* 16, 3 (1981): 299-316.

440a. Ellis, W.E. "Tenement House Reform: Another Episode in Kentucky Progressivism." *Filson Club History Quarterly* 55, 4 (1981): 375-382.

441. Fairclough, A. "The Public Utilities Industry in New Orleans: A Study in Capital Labor and Government, 1894-1929." *Journal of the Louisiana Historical Association and the Louisiana Historical Society* 22, 1 (1981): 45-66.

442. Fisher, J.A., and S.O. Park. "Locational Dynamics of Manufacturing in the Atlanta Metropolitan Region, 1968-1976." *Southeastern Geographer* 20, 2 (1980): 100-119.

443. Fisher, R. "Hispanics in Atlanta." *Atlanta Historical Journal* 24 (Summer 1980): 31-38.

444. Flowerdew, R. "Spatial Patterns of Residential Segregation in a Southern City." *Journal of American Studies* 13, 1 (1979): 93-109.

445. Frank, F.S. "Nashville Jewry During the Civil War." *Tennessee Historical Quarterly* 39, 3 (1980): 310-322.

446. Freyer, T.A. "Politics and Law in the Little Rock Crisis, 1954-1957." *Arkansas Historical Quarterly* 40, 3 (1981): 195-219.

447. Gavins, R. "Urbanization and Segregation: Black Leadership Patterns in Richmond, Virginia, 1900-1920." *South Atlantic Quarterly* 79, 3 (1980): 252-273.

448. George, P.S. "Passage to the New Eden: Tourism in Miami from Flagler through Everst G. Sewell." *Florida Historical Quarterly* 59, 4 (1980): 440-463.

449. Gildrie, R.P. "Lynch Law and the Great Clarksville Fire of 1878: Social Order in a New South Town." *Tennessee Historical Quarterly* 42, 1 (1983): 58-75.

450. Goldfield, D.R. *Cotton Fields and Skyscrapers: Southern City and Region, 1607-1980.* Baton Rouge, LA: Louisiana State University Press, 1982.

451. Goldfield, D.R. "The Urban South: A Regional Framework." *American Historical Review* 86, 5 (1981): 1009-1034.

452. Grable, S. "Applying Urban History to City Planning: A Case Study in Atlanta." *Public Historian* 1, 4 (1979): 45-59.

453. Green, R.D. "Urban Industry, Black Resistance, and
      Racial Restriction in the Antebellum South: A General
      Model and a Case Study in Urban Virginia." *Journal
      of Economic History* 41, 1 (1981): 189-191.

453a. Guethlein, C. "Women in Louisville: Moving toward
      Equal Rights." *Filson Club History Quarterly* 55, 2
      (1981): 151-178.

454. Haas, E.F. "John Fitzpatrick and Political Continuity
      in New Orleans, 1896-1899." *Journal of the Louisiana
      Historical Association and the Louisiana Historical
      Society* 22, 1 (1981): 7-30.

455. Harris, R. "Richmond: Former Confederate Capital Finally
      Falls to Blacks." *Esquire* 35, 8 (1980): 44-52.

456. Henderson, W.D. "A Great Deal of Enterprise: The
      Petersburg Cotton Mills in the Nineteenth Century."
      Virginia Cavalcade 30, 4 (1981): 176-185.

457. Holmes, R.A. "The University and Politics in Atlanta:
      A Case Study of the Atlanta University Center."
      Atlanta Historical Journal 25 (Spring 1981): 49-66.

458. Hunter, F. *Community Power Succession: Atlanta's Policy-
      Makers Revisited.* Chapel Hill, NC: University of
      North Carolina Press, 1980.

459. Hurst, H.W. "Decline and Renewal: Alexandria before
      the Civil War." *Virginia Cavalcade* 31, 1 (1981):
      32-37.

460. Huth, T. "Should Charleston Go New South?" *Historic
      Preservation* 31, 3 (1979): 32-38.

461. Jackson, P.I., and L. Carroll. "Race and the War on
      Crime: The Sociopolitical Determinants of Municipal
      Police Expenditures in 90 Southern U.S. Cities."
      American Sociological Review 46, 3 (1981): 290-305.

462. Jarzombek, M. "The Memphis-South Memphis Conflict,
      1826-1850." *Tennessee Historical Quarterly* 41, 1
      (1982): 23-36.

463. Johnson, C.W., and C.O. Jackson. *City Behind a Fence:
      Oak Ridge, Tennessee, 1942-1946.* Knoxville, TN:
      University of Tennessee Press, 1981.

464. Johnson, K.R. "Slavery and Racism in Florence, Alabama,
      1841-1862." *Civil War History* 27, 2 (1981): 155-171.

465. Johnson, M.P. "Planters and Patriarchy: Charleston,
      1800-1860." *Journal of Southern History* 46, 1 (1980):
      45-72.

466. Johnson, W.B. "Free Blacks in Antebellum Savannah: An Economic Profile." *Georgia Historical Quarterly* 64, 4 (1980): 418-431.

467. Jones, J.B. "Mose the Bower B'hoy and the Nashville Volunteer Fire Department, 1840-1860." *Tennessee Historical Quarterly* 40 (Summer 1981): 170-181.

468. Jordan, L.W. "Police and Politics: Charleston in the Gilded Age 1880-1900." *South Carolina Historical Magazine* 81, 1 (1980): 35-50.

469. ———. "Education for Community: C.G. Memminger and the Origination of Common Schools in Antebellum Charleston." *South Carolina Historical Magazine* 83, 2 (1982): 95-115.

470. Jud, G.D. "The Effects of Zoning on Single-Family Residential Property Values: Charlotte, North Carolina." *Land Economics* 56, 2 (1980): 142-154.

471. Kalmar, K.L. "Southern Black Elites and the New Deal: A Case Study of Savannah, Georgia." *Georgia Historical Quarterly* 65, 4 (1981): 341-355.

472. King, G.W. "The Emergence of Florence, South Carolina, 1853-1890." *South Carolina Historical Magazine* 82, 3 (1981): 197-209.

473. King, S.B. *Darien: The Death and Rebirth of a Southern Town.* Macon, GA: Mercer University Press, 1981.

474. Koch, Mary L. "Letters from La Grange: The Correspondence of Caroline Haralson and Her Daughter." *Georgia Historical Quarterly* 66, 1 (1982): 33-46.

475. Koebel, C.T., et al. *Housing Prices and Mortgage Lending: An Analysis of Housing Sales in Louisville.* Louisville, KY: University of Louisville Urban Studies Center, 1981.

476. Kurtz, W.G., and A. Laurie. "The Kurtz Chronicles of Early Atlanta." *Atlanta Historical Journal* 26 (Spring 1982): 5-32.

477. Lachance, P.F. "Intermarriage and French Cultural Persistence in Late Spanish and Early American New Orleans." *Social History* 15, 29 (1982): 47-82.

478. ———. "The Urban South in Historical Perspective." *Canadian Review of American Studies* 13, 1 (1982): 53-60.

479. Lack, P.D. "Law and Disorder in Confederate Atlanta." *Georgia Historical Quarterly* 66, 2 (1982): 171-195.

480. Latimer, M. "Black Political Representation in the
     Southern Cities: Election Systems and Other Causal
     Variables." *Urban Affairs Quarterly* 15, 1 (1979):
     65-86.

481. Lawrence, D.M. *North Carolina Cities: An Introduction*.
     Chapel Hill, NC: University of North Carolina Press,
     1980.

482. Laws, K.J. "The Origin of the Street Grid in Atlanta's
     Urban Core." *Southeastern Geographer* 19, 2 (1979):
     69-79.

483. Lawson, S.F., and R.P. Ingalls. "Tampa Bay History:
     An Experiment in Public History." *Public Historian*
     3, 2 (1981): 53-62.

484. Lebsock, S. "Free Black Women and the Question of
     Matriarchy: Petersburg, Virginia, 1784-1820." *Feminist Studies* 8, 2 (1982): 271-292.

485. Lefever, H.G. "Prostitution, Politics and Religion:
     The Crusade against Vice in Atlanta in 1912." *Atlanta
     Historical Journal* 24 (Spring 1980): 7-29.

486. Luebke, P. "Activists and Asphalt: A Successful Anti-
     expressway Movement in a New South City." *Human Organization* 40, 3 (1981): 256-276.

487. Martin, C.H. "Southern Labor Relations in Transition:
     Godsden, Alabama, 1930-1943." *Journal of Southern
     History* 47, 4 (1981): 545-568.

488. McCoy, C.B., and V.M. Watkins. "Drug Use among Urban
     Ethnic Youth: Appalachian and Other Comparisons."
     Youth and Society 12, 1 (1980): 83-106.

489. McGovern, J.R. "Pensacola, Florida: A Military City
     in the New South." *Florida Historical Quarterly*
     59, 1 (1980): 24-41.

490. Megna, R. *More Than a Memory: Little Rock's Historic
     Quapaw Quarter*. Little Rock, AK: Rose Publishing,
     1981.

490a. Messmer, C.K. "The End of an Era: Louisville in 1865."
     Filson Club Historical Quarterly 54, 3 (1980): 239-
     271.

491. Miller, Zane L. *Suburb: Neighborhood and Community in
     Forest Park, 1935-1976*. Knoxville, TN: University
     of Tennessee Press, 1981.

492. Mitchell, M.V., and R. Holladay. "A Strategy for the

1980's: Memphis' Experience in Housing Inspection." *Journal of Housing* 37, 1 (1980): 21-26.

493. Mohl, R.A. "Forty Years of Economic Change in the Miami Area." *Florida Environmental and Urban Issues* 9, 4 (1982): 8-14.

494. ———. "Race, Ethnicity and Urban Politics in the Miami Metropolitan Area." *Florida Environmental and Urban Issues* 9, 3 (1982): 1-7.

495. Moore, J.H. "Appomatox: Profile of a Mid-Nineteenth-Century Community." *Virginia Magazine of History and Biography* 88, 4 (1980): 478-491.

496. ———. "Local and State Government of Antebellum Mississippi." *Journal of Mississippi History* 44 (1982): 104-134.

497. ———. "The Norfolk Riot: 16 April 1866." *Virginia Magazine of History and Biography* 90, 2 (1982): 155-164.

498. Moore, J.W. "Low Country in Economic Transition: Charleston since 1865." *South Carolina Historical Magazine* 80, 2 (1979): 156-172.

499. Mormino, G.R. "Tampa and the New Urban South: The Wight Strike of 1899." *Florida Historical Quarterly* 60, 3 (1982): 337-356.

500. Munn, Robert F. "Development of Model Towns in the Bituminous Coal Fields." *West Virginia History* 40, 3 (1979): 243-254.

501. Murray, M.A. "Energy Use Patterns in Georgia as They Relate to Population, Urbanization and Quality-of-Life." *Southeastern Geographer* 22, 1 (1982): 20-31.

502. Nichols, C.G. "Pulaski Heights: Early Suburban Development in Little Rock Arkansas." *Arkansas Historical Quarterly* 41, 2 (1982): 129-145.

503. O'Brien, J.T. "Factory, Church, and Community: Blacks in Antebellum Richmond." *The Journal of Southern History* 44, 4 (1978): 509-536.

504. ———. "Reconstruction in Richmond: White Restoration and Black Protest, April-June 1985." *Virginia Magazine of History and Biography* 89, 3 (1981): 259-281.

505. O'Loughlin, J., and D.C. Munski. "Housing Rehabilitation in the Inner City: A Comparison of Two Neighborhoods in New Orleans." *Economic Geography* 55, 1 (1979): 52-70.

506. O'Mara, J. "Town Founding in Seventeenth-Century North
      America: Jamestown in Virginia." *Journal of Historical
      Geography* 8 (January 1982): 1-11.

507. Pease, J.H., and W.H. Pease. "The Economics and Poli-
      tics of Charleston's Nullification Crises." *Journal
      of Southern History* 47, 3 (1981): 335-362.

508. Plank, D.N., and P.E. Peterson. "Does Urban Reform
      Imply Class Conflict? The Case of Atlanta's Schools."
      History of Education Quarterly 23, 2 (1983): 151-174.

509. Platt, H.L. *City Building in the New South.* Phila-
      delphia, PA: Temple University Press, 1983.

510. Preisser, T.M. "Alexandria and the Evolution of the
      Northern Virginia Economy, 1749-1776." *Virginia
      Magazine of History and Biography* 89, 3 (1981): 282-
      293.

511. Preston, H.L. *Automobile Age Atlanta: The Making of
      a Southern Metropolis, 1900-1935.* Athens, GA: Uni-
      versity of Georgia Press, 1980.

512. Rabinowitz, H.N. *Race Relations in the Urban South,
      1865-1890.* Urbana, IL: University of Illinois Press,
      1980.

513. Reynolds, T.S. "Cisterns and Fires: Shreveport, Louisi-
      ana, as a Case Study of the Emergence of Public Water
      Supply Systems in the South." *Louisiana History* 22, 4
      (1981): 337-367.

514. Rice, B.R. "The Battle of Buckhead: The Plan of Im-
      provement and Atlanta's Last Big Annexation."
      Atlanta Historical Journal 25 (Winter 1981): 5-22.

515. Robbins, P. "Alas, Memphis." *American History Illus-
      trated* 16 (January 1982): 39-46.

516. Roper, J.H. "Waterford: Preserving Open Space Is the
      Challenge Facing This Remarkable Virginia Community."
      American Preservation 2, 5 (1979): 47-57.

517. Ross, E.L. *Factors in Residence Patterns among Latin
      Americans in New Orleans, Louisiana.* New York: Arno
      Press, 1980.

518. Russell, J.F., and W.H. Berensten. "Urban Regions in
      Georgia: 1964-1979." *Southeastern Geographer* 21, 2
      (1981): 84-107.

519. Sander, R.A. "Municipal Markets in New Orleans."
      Journal of Cultural Geography 2, 1 (1981): 82-95.

520. Sears, J.N. *The First One Hundred Years of Town Planning in Georgia*. Atlanta, GA: Cherokee Publishing Co., 1979.

521. Serow, W.J., et al. "An Analysis of Population Growth Rates within Typological Categories of Small Southern Cities." *Review of Regional Studies* 10, 1 (1980): 29-47.

521a. Share, A.J. *Cities in the Commonwealth: Two Centuries of Urban Life in Kentucky*. Lexington, KY: University Press of Kentucky, 1982.

522. Sheldon, M.B. "Black-White Relations in Richmond, Virginia, 1782-1820." *The Journal of Southern History* 45, 1 (1979): 27-44.

523. Shiras, G. "Eureka Springs: One Stop and You'll Never Want to Leave This Town in the Arkansas Ozarks." *American Preservation* 2, 4 (1979): 44-58.

524. Silver, C. "A New Look at Old South Urbanization: The Irish Worker in Charleston, South Carolina, 1840-1860." *South Atlantic Urban Studies* 3 (1979): 141-172.

525. ————. "The Ordeal of City Planning in Postwar Richmond, Virginia." *Journal of Urban History* 10, 1 (1983): 33-60.

526. Smith, J.R. "The Day of Atlanta's Big Fire." *Atlanta Historical Journal* 24 (Fall 1980): 58-66.

526a. Smith, P.B. "Conserving Charleston's Architectural Heritage." *Town Planning Review* 50, 4 (1979): 459-476.

527. Sommers, R. *Richmond Redeemed: The Siege at Petersburg*. Garden City, NY: Doubleday and Co., 1981.

527a. Stephanides, M. "The Greek Community of Louisville." *Filson Club Historical Quarterly* 55, 1 (1981): 5-27.

528. Summerville, J. "The City and the Slum: Black Bottom in the Development of South Nashville." *Tennessee Historical Quarterly* 40, 2 (1981): 182-192.

529. Svara, J.H. "Attitudes Toward City Government and Preference for District Elections." *South Atlantic Urban Studies* 3 (1979): 67-84.

530. Tansey, R. "Out-of-State Free Blacks in Late Antebellum New Orleans." *Louisiana History* 22, 4 (1981): 369-386.

531. ———. "Prostitution and Politics in Antebellum New
      Orleans." *Southern Studies* 18, 4 (1979): 449-479.

532. Thomason, P. "The Men's Quarter of Downtown Nashville."
      Tennessee Historical Quarterly 41, 1 (1982): 48-66.

533. Thorn, J. "The Bell Factory: Early Pride of Huntsville."
      Alabama Review 32, 1 (1979): 28-37.

534. Thornton, J.M. "Challenge and Response in the Mont-
      gomery Bus Boycott of 1955-1956." *Alabam Review*
      33, 3 (1980): 163-235.

535. Vandel, G. "The Origins of the New Orleans Riot of
      1866, Revisited." *Louisiana History* 22, 2 (1981):
      135-165.

536. Vyhnanek, L. "Muggles, Inchy, and Mud: Illegal Drugs
      in New Orleans during the 1920's." *Louisiana History*
      22, 3 (1981): 253-279.

537. Walls, E. "Observations on the New Orleans Yellow-
      Fever Epidemic, 1878." *Louisiana History* 23, 1
      (1982): 60-67.

538. Ward, J.C. "The Election of 1880 and Its Impact on
      Atlanta." *Atlanta Historical Journal* 25 (Spring 1981):
      5-15.

539. Watts, E.J. *The Social Bases of City Politics: Atlanta,
      1865-1903.* Westport, CT: Greenwood Press, 1978.

540. Wheeler, J.O. "Effects of Geographical Scale on Loca-
      tion Decisions in Manufacturing: The Atlanta Example."
      Economic Geography 57, 2 (1982): 134-145.

541. ———, and S.O. Park. "Intrametropolitan Locational
      Changes in Manufacturing: The Atlanta Metropolitan
      Area, 1958-1976." *Southeastern Geographer* 21, 1
      (1981): 16-25.

542. Williams, B.S. "Anti-Semitism and Shreveport, Louisiana:
      The Situation in the 1920's." *Louisiana History* 21, 4
      (1980): 387-398.

543. Wilson, K.L., and A. Portes. "Immigrant Enclaves: An
      Analysis of the Labor Market Experiences of Cubans in
      Miami." *American Journal of Sociology* 86, 2 (1980):
      295-319.

544. Winsberg, M.D. "Changing Distribution of the Black
      Population: Florida Cities 1970-1980." *Urban Affairs
      Quarterly* 18, 3 (1983): 361-370.

545. ———. "Housing Segregation of a Predominantly Middle-
      Class Population: Residential Patterns Developed by

the Cuban Immigration into Miami, 1950-1974." *American Journal of Economics and Sociology* 38, 4 (1979): 403-418.

546. Wolf, E.C. "Wheeling's German Singing Societies." *West Virginia History* 42, 1/2 (1980): 1-56.

547. Wolfe, M.R. "Changing the Face of Southern Appalachia." *American Planning Association Journal* 47, 3 (1981): 252-265.

548. Woods, R. "Ethnic Segregation in Birmingham in the 1960's and 1970's." *Ethnic and Racial Studies* 2, 4 (1979): 455-476.

549. Wright, M.F. "Vickoburg and the Trans-Mississippi Supply Line (1861-1863)." *Journal of Mississippi History* 43, 4 (1981): 210-225.

550. Abbott, C. "Necessary Adjuncts to Its Growth: The Railroad Suburbs of Chicago, 1854-1875." *Journal of Illinois State Historical Society* 73, 2 (1980): 117-131.

551. Alanen, A.R., and T.J. Peltin. "Kohler, Wisconsin: Planning and Paternalism in a Model Industrial Village." *Journal of the American Institute of Planners* 44, 2 (1978): 145-159.

552. Allen, R.C. "Motion Picture Exhibition in Manhattan 1906-1912: Beyond the Nickelodeon." *Cinema Journal* 18, 2 (1979): 2-15.

553. Anderson, E. "Prostitution and Social Justice: Chicago 1910-1915." *Social Service Review* 48, 2 (1974): 184-202.

554. Anderson, H.H. "Milwaukee Novelists Depict the Local Heritage." *Milwaukee History* 3, 1 (1980): 2-15.

555. Bardell, E.B. "The Music in the Air When Milwaukee Was Young." *Historical Messenger* 30, 4 (1974): 94-106.

556. Bardo, J.W., and J.B. Hughey. "A Second-Order Factor Analysis of Community Satisfaction in a Midwestern City." *Journal of Social Psychology* 109, 2 (1979): 231-236.

557. Baron, J.N. "Indianapolis and Beyond: Structural Model of Occupational Mobility across Generations." *American Journal of Sociology* 85, 4 (1980): 815-839.

558. Bayer, F. "Computing in the City of Milwaukee: The Data Base Geographical Information System." *Computers, Environment and Urban Systems* 7, 1 (1982): 83-88.

559. Bergesen, A. "Race Riots of 1967: An Analysis of Police Violence in Detroit and Newark." *Journal of Black Studies* 12, 3 (1982): 261-274.

560. Bergstrom, W.N. "Cigar Making in Milwaukee: The Eclipse of an Industry." *Historical Messenger* 33, 1 (1977): 25-36.

561. Bernard, A.S., and M.J. Scott. "Milwaukee Revenue Cost
     Analysis." *Urban Land* 36, 12 (1977): 16-22.

562. Berry, B.J., et al. *Chicago: Transformations of an
     Urban System*. Cambridge, MA: Ballinger Publishing
     Co., 1976.

563. Berry, D. "The Sensitivity of Dairying to Urbanization:
     A Study of Northeastern Illinois." *Professional
     Geographer* 31, 2 (1979): 170-176.

564. Best, J. "Keeping the Peace in St. Paul: Crime, Vice
     and Police Work, 1869-1874." *Minnesota History* 47, 6
     (1981): 240-248.

565. Black, J.F. *Hyde Park Houses: An Informal History 1856-
     1910*. Chicago, IL: University of Chicago Press, 1978.

566. Blair, J.P., and R.S. Edari. *Milwaukee's Economy:
     Market Forces, Community Problems, and Federal Poli-
     tics*. Chicago, IL: Federal Reserve Bank, 1978.

567. Blocker, J.S. "Market Integration, Urban Growth and
     Economic Changes in an Ohio County, 1850-1900."
     Ohio History 90, 4 (1981): 298-316.

568. Bluestone, D.M. *Cleveland, An Inventory of Historic
     Engineering and Industrial Sites*. Washington, DC:
     Historic American Engineering Record, 1978.

569. Bowly, D. *The Poorhouse: Subsidized Housing in Chicago,
     1895-1976*. Carbondale, IL: Southern Illinois Univer-
     sity Press, 1978.

570. Bowron, B.R. *Henry B. Fuller of Chicago: The Ordeal of
     a Genteel Realist in Ungenteel America*. Westport,
     CT: Greenwood Press, 1974.

571. Boyd, H. "What's Happening in Murder City." *Progressive*
     45, 9 (1981): 38-42.

572. Bradbury, K.L., et al. *Futures for a Declining City:
     Simulations for the Cleveland Area*. New York: Academic
     Press, 1981.

573. Bremer, S.H. "Lost Continuities: Alternative Urban
     Visions in Chicago Novels, 1890-1915." *Soundings:
     An Interdisciplinary Journal* 64, 3 (1981): 29-51.

574. Brook, A. "Gary, Indiana: Steeltown Extraordinary."
     Journal of American Studies 9, 1 (1975): 35-55.

575. Brown, A., and L. Dorsett. *A History of Kansas City,
     Missouri*. Boulder, CO: Pruett Publishing Co., 1978.

576. Brown, T.J. "The Age of Ambition in Quincy, Illinois."
      Journal of the Illinois State Historical Society 75,
      4 (1982): 242-262.

577. Buettinger, C. "Economic Inequality in Early Chicago
      1840-1850." *Journal of Social History* 11, 3 (1978):
      413-418.

578. Burg, D.F. *Chicago's White City of 1893*. Lexington,
      KY: University Press of Kentucky, 1976.

579. Cain, L.P. *Sanitation Strategy for a Lakefront Metropo-
      lis: The Case of Chicago*. Dekalb, IL: Northern Illi-
      nois University Press, 1979.

580. ————. "Unfouling the Public's Nest: Chicago's Sani-
      tary Diversion of Lake Michigan Water (1881-1930)."
      Technology and Culture 15, 4 (1974): 594-613.

581. Cangi, E.C. "Patrons and Protégés: Cincinnati's First
      Generation of Women Doctors, 1875-1910." *Cincinnati
      Historical Society Bulletin* 37, 2 (1979): 89-114.

582. Cary, L.L. "The Bureau of Investigation and Radicalism
      in Toledo, Ohio 1918-1920." *Labor History* 21, 3
      (1980): 430-440.

583. Cassity, M.J. "Modernization and Social Crisis: The
      Knights of Labor and a Mid-West Community." *Journal
      of American History* 66, 1 (1979): 41-61.

584. Christensen, D. *Chicago: Chicago Public Works, a His-
      tory*. Chicago, IL: Rand McNally and Co., 1974.

585. Conn, S. "Three Talents: Roben, Nestor, and Anderson
      of the Chicago Women's Trade Union League." *Chicago
      History* 9, 4 (1980-81): 234-247.

586. Conzen, K.N. "Approaches to Early Milwaukee Community
      History." *Milwaukee History* 1, 1/2 (Spring/Summer
      1978): 4-12.

587. Conzen, M.P. "Capital Flows and the Developing Urban
      Hierarchy: State Bans Capital in Wisconsin, 1845-
      1895." *Economic Geography* 51, 4 (1975): 321-338.

588. ————., and K.N. Conzen. "Geographical Structure
      in Nineteenth-Century Urban Retailing: Milwaukee 1836-
      1890." *Journal of Historical Geography* 5, 1 (1979):
      45-66.

589. Cronon, W.J. "To Be the Central City: Chicago, 1848-
      1857." *Chicago History* 10, 3 (1981): 130-140.

590. Curran, D.J. *Metropolitan Financing: The Milwaukee Experience, 1920-1970.* Madison, WI: University of Wisconsin Press, 1973.

591. Curry, R.P. "From Instability to Stability: The Courter Heinhold Influence." *The Cincinnati Historical Society Bulletin* 39, 3 (1981): 159-174.

592. Cutler, I. *Chicago: Metropolis of the Mid-Continent.* Chicago, IL: Geographic Society of Chicago, 1973.

593. Danek, E. *Cedar Rapids (Iowa): Tall Corn and High Technology.* Woodland Hills, CA: Windsor Publications, 1980.

594. Dannenbaum, J. "Immigrants and Temperance: Ethnocultural Conflict in Cincinnati 1845-1860." *Ohio History* 87, 2 (1978): 125-139.

595. Davis, R.L. "Community and Conflict in Pioneer Saint Louis, Missouri." *Western Historical Quarterly* 10, 3 (1979): 337-356.

596. Demeter, C.S. "The Joe Louis Arena Site: A Material Cultural View of Nineteenth-Century Detroit." *Detroit Perspective* 4, 2 (1980): 143-169.

597. DeMuth, J. *Small-Town Chicago: The Comic Perspective of Finley Peter Dunne, George Ade, and Ring Lardner.* Port Washington, NY: Kennikat Press, 1980.

598. Dow, L.M. "High Weeds in Detroit: The Irregular Economy among a Network of Appalachian Migrants." *Urban Anthropology* 6, 2 (1977): 111-128.

599. Dragos, S.F. "Privately Funded Mechanisms for Milwaukee Redevelopment." *Urban Land* 36, 1 (1977): 14-23.

600. Drury, J. *Old Chicago Houses.* Chicago, IL: University of Chicago Press, 1975.

601. Eaton, L.K. "Warehouses and Warehouse Districts in Mid-American Cities." *Urban History Review* 11, 1 (1982): 17-26.

602. Eckstein, L., and S. Plattner. "Ethnicity and Occupations in Soulard Farmers Market, St. Louis, Missouri." *Urban Anthropology* 7, 4 (1979): 361-372.

603. Ehrlich, G. "Partnership Practice and the Professionalization of Architecture in Kansas City, Missouri." *Missouri Historical Review* 74, 4 (1980): 458-480.

604. Eigel, J.C. "Surviving Prohibition in Milwaukee." *Historical Messenger* 33, 4 (1977): 118-124.

605. Entry deleted.

606. Engelen, R.E. "What Is the Future for Downtown Re-
      tailing in Middle America? A Middle Market Malaise."
      Urban Land 38, 9 (1979): 5-11.

607. Erickson, R.A., and M. Wasylenko. "Firm Relocation and
      Site Selection in Suburban Municipalities." *Journal
      of Urban Economics* 8, 1 (1980): 69-85.

608. Ewalt, D.H., and G.R. Kremer. "The Historian as Preser-
      vationist: a Missouri Case Study." *Public Historian*
      4, 4 (1981): 5-22.

609. Ewen, L.A. *Corporate Power and Urban Crisis in Detroit.*
      Princeton, NJ: Princeton University Press, 1978.

610. Fairbanks, R.B. "Housing the City: The Better Housing
      League and Cincinnati 1916-1939." *Ohio History* 39, 2
      (1980): 157-180.

611. Farley, J.E. "Metropolitan Housing Segregation in 1980:
      The St. Louis Case." *Urban Affairs Quarterly* 18, 3
      (1983): 347-360.

612. Fine, H.D. "The Koreshan Unity: The Chicago Years of
      a Utopian Community." *Journal of the Illinois State
      Historical Society* 67, 2 (1975): 213-228.

613. Fine, S. *Frank Murphy: The Detroit Years.* Ann Arbor,
      MI: University of Michigan Press, 1975.

614. Finkelman, P. "Class and Culture in Late Nineteenth-
      Century Chicago: The Founding of the Newberry Library."
      American Studies 16, 1 (1975): 5-23.

615. Floyd, C. "A Sense of Pride: Cincinnati Neighborhoods
      Document Their Histories." *History News* 36, 10
      (1981): 13-14.

616. Foley, William E. "St. Louis, the First Hundred Years."
      The Bulletin of the Missouri Historical Society 34,
      4, pt. 1 (1978): 187-199.

617. Frank, M.L. "In North Minneapolis: Sawmill City Boy-
      hood." *Minnesota History* 47, 4 (1980): 141-153.

618. Fry, M.C. "Women on the Ohio Frontier: The Marietta
      Area." *Ohio History* 90, 1 (1981): 55-73.

619. Gayle, M. "A Heritage Forgotten: Chicago's First Cast
      Iron Front Buildings." *Chicago History* 7, 2 (1978):
      98-108.

620. Georgakas, D., and M. Surkin. *Detroit: I Do Mind Dying.*
      New York: St. Martin's Press, 1975.

621. Ghent, J.M., and F.C. Jaher. "The Chicago Business
      Elite: 1830-1930." *Business History Review* 50, 3
      (1976): 288-328.

622. Gjeide, J. "The Effect of Community on Migration:
      Three Minnesota Townships 1885-1905." *Journal of
      Historical Geography* 5, 4 (1979): 403-422.

623. Goldberg, R.A. "The Ku Klux Klan in Madison, 1922-
      1927." *Wisconsin Magazine of History* 57, 1 (1974):
      31-44.

624. Goniery, D. "Movie Exhibition in Milwaukee, 1906-1947:
      A Short History." *Milwaukee History* 2, 1 (1979):
      8-17.

625. Goodwin, C. *The Oak Park Strategy: Community Control
      of Racial Change*. Chicago, IL: University of Chicago
      Press, 1979.

626. Gorton, T. "Detroit Reborn." *Planning* 45, 7 (1979):
      14-19.

627. Green, H.L., et al. "Where Are Store Locations Good?
      The Case of the National Tea Company in Detroit."
      Professional Geographer 30, 2 (1978): 162-167.

628. Entry deleted.

629. Hagensick, A.C., and B. Rasmussen. "Wisconsin's Local
      Affairs Agency: Its Rise and Demise." *National Civic
      Review* 70, 11 (1981): 589-593.

630. Haller, M.H. "Historical Roots of Police Behavior:
      Chicago 1890-1925." *Law and Society Review* 10, 1
      (1975): 303-324.

631. Havira, B.S. "Managing Industrial and Social Tensions
      in a Rural Setting: Women Silk Workers in Belding,
      Michigan, 1885-1932." *Michigan Academician* 13, 3
      (1981): 257-273.

632. Heatley, H. "The Wardwell House: A Legacy of Old Grosse
      Pointe." *Detroit Perspective* 5, 1 (1981): 26-44.

633. Heise, K. *Is There Only One Chicago?* Richmond, VA:
      Westover Publishing Company, 1973.

634. Henderson, J.P. "The History of Thought in the Develop-
      ment of the Chicago Paradigm." *Journal of Economic
      Issues* 10, 1 (1976): 127-148.

635. Hokanson, N.M. "The Foreign Language Division of the
      Chicago Liberty Loan Campaign." *Journal of the Illinois
      State Historical Society* 67, 4 (1974): 429-439.

636. Holli, M.G. "Detroit Today: Locked into the Past."
Midwest Quarterly 19, 3 (1978): 251-259.

637. Holt, Glen E. "The Future of St. Louis: Another Look
Ahead." *The Bulletin of the Missouri Historical
Society* 34, 4, pt. 1 (1978): 211-217.

638. Horowitz, A.J. "Developable Area and Trip Distribution
in Residential Location: A Detroit Case Study."
Journal of Regional Science 18, 4 (1978): 429-446.

639. Horowitz, H.C. *Culture and the City: Cultural Philanthropy in Chicago from the 1880's to 1917.* Lexington,
KY: University Press of Kentucky, 1976.

640. Horowitz, R. "Adult Delinquent Gangs in a Chicago Community: Masked Intimacy and Marginality." *Urban Life*
11, 1 (1982): 3-26.

641. Hunter, A. *Symbolic Communities: The Persistent Change
of Chicago's Local Communities.* Chicago, IL: Univer-
siry of Chicago Press, 1982.

642. ————. "Why Chicago? The Rise of the Chicago School
of Urban Social Science." *American Behavioral Scientist* 24, 2 (1980): 215-227.

643. Hurt, R.D. "Pork and Porkopolis." *Cincinnati Historical Society Bulletin* 40, 2 (1982): 191-212.

644. Jebsen, H. "Preserving Suburban Identity in an Expanding Metropolis: The Case of Blue Island, Illinois."
Old Northwest 7, 2 (1981): 127-145.

645. Johnson, W.C., and J.J. Harrigan. "Innovation by In-
crements: The Twin Cities as a Case Study in Metropolitan Reform." *Western Political Quarterly* 31, 2
(1978): 206-218.

646. ————. "Planning for Growth: The Twin Cities Approach."
National Civic Review 68, 4 (1979): 189-193.

647. Keating, S.L. "The French Farmers of Grosse Pointe:
The Jean Baptiste Revard Family." *Detroit Perspectives*
5, 2 (1981): 72-84.

648. Key, D. "Milwaukee's Art of the Depression Era." *Historical Messenger* 31, 2 (1975): 38-50.

649. Klang, Y. "Recent Changes in the Distribution of Urban
Poverty in Chicago." *Professional Geographer* 28, 1
(1976): 57-61.

650. Kocolowski, Gary P. "Stabilizing Migration to Louisville
and Cincinnati, 1865-1906." *Cincinnati Historical
Society Bulletin* 37, 1 (1979): 23-47.

651. Konet, J. "Milwaukee's Nickelodeon Era, 1906-1915."
     Milwaukee History 2, 1 (1979): 2-7.

652. Koprowski-Kraut, G. "The Depression's Effects on a
     Milwaukee Family." *Milwaukee History* 3, 3 (1980):
     84-92.

653. Kozllowski, P.J. *Business Conditions in Michigan Metro-
     politan Areas*. Kalamazoo, MI: W.E. Upjohn Institute
     for Employment, 1979.

654. Kremer, G.R., and T.E. Gage. "The Prison against the
     Town: Jefferson City and the Penitentiary in the
     Nineteenth Century." *Missouri Historical Review*
     74, 4 (1980): 414-432.

655. Kuhm, H.W. "Dr. Robert J. Faries: Pioneer Milwaukee
     Dentist." *Historical Messenger* 32, 1 (1976): 2-10.

656. Lane, J.B. *City of the Century: A History of Gary,
     Indiana*. Bloomington, IN: Indiana University Press,
     1978.

657. Lauer, J.C., and R.H. Lauer. "St. Louis and the 1880
     Census: The Shock of Collective Failure." *Missouri
     Historical Review* 76, 2 (1982): 151-163.

658. Leonard, C., and J. Walliman. "Prostitution and Changing
     Morality in the Frontier Cattle Towns of Kansas."
     Kansas History 2, 1 (1979): 34-54.

659. Levathes, L. "Milwaukee: More Than Beer." *National
     Geographic* 158, 8 (1980): 180-232.

660. Linden, B.M. "Inns to Hotels in Cincinnati." *Cincin-
     nati Historical Society Bulletin* 39, 3 (1981): 127-
     152.

661. Littlejohn, E.L. "Law and Police Misconduct: The Crisis
     of the Wounded: A History of Police Misconduct in
     Detroit." *Journal of Urban Law* 58, 2 (1981): 173-
     220.

662. Longstreet, S. *Chicago: An Intimate Portrait of People,
     Pleasures, and Power 1860-1919*. New York: David MacKay
     Company, 1973.

663. Lorence, J.J. "The Milwaukee Connection: The Urban-
     Rural Link in Wisconsin Socialism; 1910-1920." *Mil-
     waukee History* 3, 4 (1980): 102-111.

664. Lurie, J. *The Chicago Board of Trade, 1859-1908: The
     Dynamics of Self-Regulation*. Urbana, IL: University
     of Illinois Press, 1979.

665. Lynch, R. "Public Works and Public History--Kansas City and Beyond." *Public Historian* 1, 3 (1979): 77-82.

666. Mandell, L. *Industrial Location Decisions: Detroit Compared with Atlanta and Chicago.* London: Martin Robertson, 1976.

667. Mapes, L., and A. Tavis. *A Pictorial History of Grand Rapids.* Grand Rapids, MI: Kregel, 1976.

668. Marcus, A.I. "Social Evils and the Origin of Municipal Services in Cincinnati." *Journal of American Studies* 22, 2 (1981): 23-40.

669. ———. "The Strange Career of Municipal Health Initiatives: Cincinnati and City Government in the Early Nineteenth Century." *Journal of Urban History* 7, 1 (1980): 3-29.

670. Margulis, H.L. "Housing Mobility in Cleveland and Its Suburbs, 1975-1980." *Geographical Review* 72, 1 (1982): 36-49.

671. Mayer, H.M. "The Launching of Chicago: The Situation and the Site." *Chicago History* 9, 2 (1980): 68-79.

672. ———. "Urban Geography and Chicago in Retrospect." *Annals of the Association of American Geographers* 69, 1 (1979): 114-117.

673. Mayer, J.A. "Relief Systems and Social Controls: The Case of Chicago, 1890-1910." *Old Northwest* 6, 3 (1980): 217-244.

674. McLear, P.E. "The Galena and Chicago Union Railroad: A Symbol of Chicago's Economic Maturity." *Journal of the Illinois State Historical Society* 73, 1 (1980): 17-26.

675. ———. "Land Speculators and Urban Regional Development: Chicago in the 1830's." *Old Northwest* 6, 2 (1980): 137-151.

676. ———. "Wilhelm Butler Ogden: A Chicago Promoter in the Speculative Era and the Panic of 1837." *Journal of the Illinois State Historical Society* 70, 4 (1977): 283-291.

677. McNabb, G. *A Short History of Marine City, Michigan.* Marine City, MI: Marine City Rotary Club, 1980.

678. Meehan, P.J. "Urban Design Criteria for Small Town Central Business Districts: Midwest, USA." *Ekistics* 49, 297 (1982): 433-441.

679. Melvin, P.M. "Make Milwaukee Safe for Babies: The
     Child Welfare Commission and the Development of
     Urban Health Centers, 1911-1912." *Journal of the
     West* 7, 2 (1978): 83-93.

680. Entry deleted.

681. Miller, Z.L. *Suburb: Neighborhood and Community in
     Forest Park, Ohio 1935-1936*. Knoxville, TN: Univer-
     sity of Tennessee Press, 1981.

682. ———, and G. Giglierano. "Downtown Housing: Changing
     Plans and Perspectives, 1948-1980." *Cincinnati His-
     torical Society Bulletin* 40, 2 (1982): 167-190.

683. Mohl, R.A. "The Great Steel Strike of 1919 in Gary,
     Indiana: Working Class Radicalism or Trade Union
     Militancy?" *Mid-America* 63, 1 (1981): 36-52.

684. Morgan, B.B. "Goal Setting by Officials: The Kansas
     City Experience." *National Civic Review* 71, 6 (1982):
     303-308.

685. O'Brien, P. "The Yeo Site: A Kansas City Hopewell
     Limited Activity Site in Northwestern Missouri and
     Some Theories." *Plains Anthropologist* 27, 95 (1982):
     37-56.

686. Ogden, M.G. "Memories of Milwaukee." *Historical Mes-
     senger* 30, 3 (1974): 81-86.

687. Ohensmann, J.R. "Urban Development in Indianapolis:
     Prospects for the Future." *School of Public and
     Environmental Affairs Review* 3, 2 (1982): 19-22.

688. Paraschos, J.N. "Mt. Auburn: Helping Residents Take
     Pride in Their Cincinnati Neighborhood." *American
     Preservation* 2, 4 (1979): 7-12.

689. Parrish, D. *Historic Architecture of Lafayette, In-
     diana*. Lafayette, IN: Purdue University Press, 1978.

690. Peirce, N. "How Can Cities Cope with Problems of the
     80's?" *Minnesota Cities* 64, 11 (1979): 4-9.

691. Pellow, D. "The New Urban Community: Mutual Relevance
     of the Social and Physical Environments." *Human
     Organization* 40, 1 (1981): 15-26.

692. Peterson, G.B., et al. *Southern Newcomers to Northern
     Cities: Work and Social Adjustment in Cleveland*.
     New York: Praeger, 1977.

693. Philpott, T.L. *The Slum and the Ghetto Neighborhood
     Deterioration and Middle-Class Reform, Chicago, 1880-
     1930*. New York: Oxford University Press, 1978.

694. Primm, J.N. *Lion of the Valley: St. Louis, Missouri*.
     Boulder, CO: Pruett Publishing Co., 1981.

695. Proress, D.L. "Banfield's Chicago Revisited: The Con-
     dition for and Social Policy Implications of the
     Transformation of a Political Machine." *Social
     Service Review* 48, 2 (1974): 184-202.

696. Prosser, D.J. "Chicago and the Bungalow Boom of the
     1920's." *Chicago History* 10, 2 (1981): 86-95.

697. Ratcliffe, J.E. *A Community in Transition: The Edge-
     water Community in Chicago*. Chicago, IL: Loyola
     University Press, 1978.

698. Reese, W.J. "Partisans of the Proletariat: The Social-
     ist Working Class and the Milwaukee Schools, 1890-
     1920." *History of Education Quarterly* 21, 1 (1981):
     3-50.

699. Rengstorf, S.R. "A Neighborhood in Transition:
     Sedamsville 1880-1950." *Cincinnati Historical
     Society Bulletin* 39, 3 (1981): 175-194.

700. Reynolds, M.S. "The City, Suburbs, and the Establish-
     ment of the Clifton Town Meeting, 1961-1964." *Cin-
     cinnati Historical Society Bulletin* 38, 1 (1980):
     7-32.

701. Reynolds, W.J. *Change on the Milwaukee Urban Fringe:
     The Case of Four Washington County Hamlets*. Milwaukee,
     WI: University of Wisconsin Press, 1980.

702. Ring, D.F. "The Temperance Movement in Milwaukee, 1872-
     1884." *Historical Messenger* 31, 4 (1975): 98-105.

703. Rose, J. "Municipal Activities and the Antitrust Laws
     after City of Lafayette." *Journal of Urban Law* 57, 3
     (1980): 483-522.

704. Rosen, G. *Decision-Making, Chicago Style*. Urbana, IL:
     University of Illinois Press, 1980.

705. Rousmaniere, K. "The Muscatine Button Workers Strike
     of 1911-1912: An Iowa Community in Conflict." *Annals
     of Iowa* 46, 4 (1982): 243-262.

706. Salvatore, N. "Railroad Workers and the Great Strike
     of 1877: The View from a Small Midwest City (Terre
     Haute)." *Labor History* 21, 4 (1980): 522-545.

707. Schlereth, T.J. "America, 1817-1919: A View of Chicago."
     Journal of American Studies 17, 2 (1976): 87-100.

708. Schmid, J.A. *Urban Vegetation: A Review and Chicago
     Case Study*. Chicago, IL: University of Chicago Press,
     1975.

709. Schneider, J.C. "Detroit and the Geography of the City: Crime, Violence, and the Police in Detroit, 1845-1875." *Journal of Urban History* 4, (Feb. 1978): 183-204.

710. ———. *Detroit and the Problem of Order: A Geography of Crime, Riot and Policing*. Lincoln, NE: University of Nebraska Press, 1980.

711. ———. "Urbanization and the Maintenance of Order: Detroit, 1824-1847." *Michigan History* 60, 5 (1976): 260-281.

712. Schneirov, R. "Chicago's Great Upheaval of 1877." *Chicago History* 9, 1 (1980): 2-17.

713. Sehr, T.J. "Three Gilded Age Suburbs of Indianapolis: Irvington, Brightwood, and Woodruff Place." *Indiana Magazine of History* 77, 4 (1981): 305-332.

714. Entry deleted.

715. Simon, R.D. *The City Building Process: Housing and Services in New Milwaukee Neighborhoods, 1880-1910*. Philadelphia, PA: Transactions of the American Philosophical Society, 1978.

716. ———. "Foundations for Industrialization, 1835-1880." *Milwaukee History* 1, 1/2 (Spring/Summer 1978): 38-56.

717. ———. "Housing and Services in an Immigrant Neighborhood: Milwaukee Ward 14." *Journal of Urban History* 2, 4 (1976): 435-458.

718. Simpson, D. *Chicago Future: An Agenda for Change*. Chicago, IL: Swallow Press, 1980.

719. Spiegel, S.A. "Affirmative Action in Cincinnati." *Cincinnati Historical Society Bulletin* 37, 2 (1979): 79-88.

720. Steiner, F. *The Politics of New Town Planning: The Newfields, Ohio Story*. Athens, Ohio: Ohio University Press, 1982.

721. Entry deleted.

722. Stickland, W.A. "Frontier Druggists of Westport, Missouri during the Westward Migration, 1848-1861." *Pharmacy History* 23, 1 (1981): 3-16.

723. Storey, D.J. "New Firm Formation, Employment Change and the Small Firm: The Case of Cleveland County." *Urban Society* 18, 3 (1981): 335-346.

724. Strassman, W.P. "Development Economics from a Chicago Perspective." *Journal of Economic Issues* 10, 1 (1976): 63-80.

725. Surber, N., and R. Bolton. "Foundation Square South: Public/Private Partnership Works in Cincinnati." *Urban Land* 41, 9 (1982): 3-5.

726. Swanstrom, T. "Tax Abatement in Cleveland." *Social Policy* 12, 3 (1982): 24-30.

727. Szuberla, G. "Three Chicago Settlements: Their Architectural Form and Social Meaning." *Journal of the Illinois State Historical Society* 70, 2 (1977): 114-129.

728. Thomas, J.C. "The Class Impacts of Budget Cutbacks: Outcomes of Retrenchment in Cincinnati." *Urban Interest* 3, 1 (1981): 51-61.

729. Todd, W.J. "Milwaukee's Lincoln Park District: A Look at Three Photomaps." *Milwaukee History* 2, 3 (1979): 58-64.

730. Triem, J. "Kansas City, Missouri: Its Historical and Architectural Significance." *Public Historian* 1, 3 (1979): 72-76.

731. Tripp, J.F. "Kansas Communities and the Birth of the Labor Problem, 1877-1883." *Kansas History* 4, 2 (1981): 114-129.

732. Walsh, M. "Business Success and Capital Availability in the New West: Milwaukee Iron Makers in the Middle Nineteenth Century." *Old Northwest* 1, 2 (1975): 159-179.

733. Watts, E.J. "Black and Blue: Afro-American Police Officers in Twentieth-Century St. Louis." *Journal of Urban History* 712 (1981): 131-168.

734. Widder, K.R. "Founding La Pointe Mission, 1825-1833." *Wisconsin Magazine of History* 64, 3 (1981): 181-201.

735. Wright, B. "The MAC Fire in St. Louis, 1914." *Missouri Historical Review* 72, 4 (1978): 424-433.

THE WEST

General

736. Einstadter, W.J. "Crime News in the Old West: Social
Control in a Northwestern Town 1887-1888." *Urban
Life* 8, 3 (1979): 317-334.

737. Luckingham, B. "The American West: An Urban Perspective." *Journal of Urban History* 8, 1 (1981): 99-105.

738. Marando, V.L. "City-County Consolidation: Reform,
Regionalism, Referenda and Requiem." *Western Political
Quarterly* 32, 4 (1970): 409-421.

739. Reps, J.W. *Cities of the American West: A History of
Frontier Urban Planning.* Princeton, NJ: Princeton
University Press, 1979.

740. ———. *The Forgotten Frontier: Urban Planning in the
American West Before 1890.* Columbia, MO: University
of Missouri Press, 1982.

Pacific Coast

741. Abbott, C. "Portland in the Pacific War: Planning from
1940-1945." *Urbanism Past and Present* 6 (1980-81):
12-24.

742. Anthony, H.A., and K.H. Anthony. "Urban Development
along the California Coast." *Ekistics* 49, 293 (1982):
160-165.

743. Banning, E.I. "U.S. Grant, Jr., a Builder of San Diego."
Journal of San Diego History 27 (Winter 1981): 1-16.

744. Bisceglia, L.R. "The McManus Welcome San Francisco,
1951." *Journal of Irish Studies* 16, 1 (1981): 6-20.

745. Blackford, M.G. "Civic Groups, Political Action, and
City Planning in Seattle, 1892-1915." *Pacific Historical Review* 49, 4 (1980): 557-580.

746. Blakely, E.J., et al. *Goal Setting for Community De-
      velopment: The Case of Yuba City, California.* Davis,
      CA: Institute of Governmental Affairs, 1978.

747. Bradley, B. *Commercial Los Angeles, 1925-1947: Photo-
      graphs from the Dick Whittington Studio.* Glendale,
      CA: Interurban Press, 1981.

748. Brambilla, R., and G. Lange. *Learning from Seattle.*
      New York: Institute for Environmental Action, 1980.

749. Brandes, R. *San Diego: An Illustrated History.* San
      Diego, CA: Howell-North Books, 1981.

750. Bruvold, W.H. "Residential Response to Urban Drought
      in Central California." *Water Resources Research* 15,
      6 (1979): 1297-1304.

751. Carroll, M. "Seattle's Madrona Town Square." *Challenge*
      10, 10 (1979): 22-27.

752. Caughey, J., and L. Caughey. *Los Angeles: Biography of
      a City.* Berkeley, CA: University of California Press,
      1977.

753. Chase, L. "Eden in the Orange Groves: Bungalows and
      Courtyard Houses of Los Angeles." *Landscape* 23, 3
      (1981): 29-36.

754. Chiles, F. "General Strike: San Francisco, 1934--An
      Historical Compilation Film Storyboard." *Labor His-
      tory* 22, 3 (1981): 430-465.

755. Chow, W.T. "Planning for Micropolitan Growth in Cali-
      fornia." *Town Planning Review* 52, 2 (1981): 184-204.

756. Cole, T. *A Short History of San Francisco.* San Fran-
      cisco, CA: Lexikes, 1981.

757. Curtis, J.R. "New Chicago in the Far West: Land Specu-
      lation in Alviso, California, 1890-1891." *California
      History* 61, 1 (1982): 36-45.

758. Davoren, W.T. "Tragedy of the San Francisco Bay Commons."
      Coastal Zone Management Journal 9, 2 (1982): 111-154.

759. Dembo, J. "John Danz and the Seattle Amusement Trades
      Strike, 1921-1935." *Pacific Northwest Quarterly* 71, 4
      (1980): 172-182.

760. Detwiler, P.M. "Public Works or Public Policy: What
      Guides New Development in California." *Urban Land*
      39, 8 (1980): 9-13.

761. Dicker, L.M. "The San Francisco Earthquake and Fire."
      California History 59, 1 (1980): 34-65.

762. Dingemans, D. "Redlining and Mortgage Lending in Sacramento." *Annals of the Association of American Geographers* 69, 2 (1979): 225-239.

763. Dixon, M. *What Happened to Fairbanks? The Effects of the Trans-Alaska Oil Pipeline on the Community of Fairbanks, Alaska.* Boulder, CO: Westview Press, 1978.

764. Eckbo, G. *Public Landscape: Six Essays on Government and Environmental Design in the San Francisco Bay Area.* Berkeley, CA: University of California. Institute of Governmental Studies, 1978.

765. Elliott, H.M. "Macrocalifornia and the Urban Gradient." *California Geographer* 21 (1981): 1-17.

766. Foster, R.H. "Wartime Trailer Housing in the San Francisco Bay Area." *Geographical Review* 70, 3 (1980): 276-290.

767. Friedman, K.J. "Urban Food Marketing in Los Angeles, 1850-1885." *Agricultural History* 54, 3 (1980): 433-445.

768. Garr, D.J. "Los Angeles and the Challenge of Growth 1839-1849." *Southern California Quarterly* 61, 2 (1979): 147-158.

769. Gleye, P. *The Architecture of Los Angeles.* Los Angeles, CA: Rosebud Books, 1981.

770. Gottdiener, M. "Disneyland: A Utopian Urban Space." *Urban Life* 11, 2 (1982): 139-162.

771. Greb, G.A. "Opening a New Frontier: San Francisco, Los Angeles, and the Panama Canal, 1900-1914." *Pacific Historical Review* 47, 3 (1978): 405-424.

772. Haas, Gilda, and Allen D. Heskin. "Community Struggles in Los Angeles." *International Journal of Urban and Regional Research* 5 (December 1981): 546-564.

773. Harr, R.D. "Fog Drip in the Bull Run Municipal Watershed, Oregon." *Water Resources Bulletin* 18, 5 (1983): 785-791.

774. Hoffman, A. "The Los Angeles Aqueduct Investigation Board of 1912: A Reappraisal." *Southern California Quarterly* 62 (Winter 1980): 329-360.

775. Hughes, R.E., and L.D. Scott. "The Founding of San Diego's Red Cross, 1898-1917." *Journal of San Diego History* 27 (Spring 1981): 115-128.

776. Kahrl, W.L. *Water and Power: The Conflict over Los Angeles' Water Supply in the Owens Valley*. Berkeley, CA: University of California Press, 1982.

777. Kaplan, S. "Remarkable Pasadena: Flair Among the Freeways." *Historic Preservation* 33, 2 (1981): 38-47.

778. Katz, H.C. "Municipal Pay Determination: The Case of San Francisco." *Industrial Relations* 18, 1 (1979): 44-58.

779. Kremer, P.C., and H.P. Oliver. "The Emerging California Shelter Crisis." *Urban Land* 40, 7 (1981): 3-8.

780. Lane, P. "Seattle: A Citizen-Watched Pot on Constant Boil." *Urban Land* 42, 3 (1983): 2-10.

781. Lawler, P. "In Boom Town Anchorage: Railroad Workers Battle Capitalist Wage Slavery." *Alaska Journal* 10, 3 (1980): 22-27.

782. Lockwood, Charles. "Rincon Hill Was San Francisco's Most Genteel Neighborhood." *California History* 63, 1 (1979): 48-62.

783. Lockwood, C. "Tourists in Gold Rush San Francisco." *California History* 59, 4 (1980/81): 314-333.

784. "Los Angeles, 1781-1981." *California History* 60, 1 (1981): 6-107.

785. Macphail, E.C. "Lydia Knapp Horton: A Liberated Woman in Early San Diego." *Journal of San Diego History* 27 (Winter 1981): 17-42.

786. Mathes, W.M. "Documents for the History of Sonoma, California 1848-1906: Calendar." *California History* 59, 3 (1980): 260-280.

787. Medler, J., and A. Mushkatel. "Urban-Rural Class Conflict in Oregon Land Use Planning." *Western Political Quarterly* 32, 3 (1979): 338-349.

788. Miller, M.J., et al. *America's Urban Capital Stock, V. 5: The Future of Oakland's Capital Plant*. Washington, DC: The Urban Institute, 1981.

789. Monkkoner, E. "Toward an Understanding of Urbanization: Drunk Arrests in Los Angeles." *Pacific Historical Review* 50, 2 (1981): 234-244.

790. Montes, G. "Balboa Park, 1909-1911, the Rise and Fall of the Olmsted Plan." *Journal of San Diego History* 28 (Winter 1982): 46-67.

791. Morgan, M.C. *Skid Row: An Informal Portrait of Seattle.* Seattle, WA: Washington University Press, 1981.

792. Moses, V. "Machines in the Garden: A Citizen's Monopoly in Riverside, 1900-1936." *California History* 61, 1 (1982): 26-35.

793. Mullins, W.H. "Self-Help in Seattle, 1931-1932: Herbert Hoover's Concept of Cooperative Individualism and the Unemployed Citizen's League." *Pacific Northwest Quarterly* 72, 1 (1981): 11-19.

794. Nielsen, D.C., and P.R. Pryde. "Providing for Rural Land in San Diego County." *Urban Land* 39, 10 (1980): 11-17.

795. Paul, R.W. "After the Gold Rush: San Francisco and Portland." *Pacific Historical Review* 51, 1 (1982): 1-21.

796. Peterson, W.J., et al. "Perception of Alcohol and Alcoholism among Alaskan Communities." *Journal of Alcohol and Drug Education* 25, 1 (1979): 31-35.

797. Polzoides, S., et al. *Courtyard Housing in Los Angeles.* Berkeley, CA: University of California Press, 1982.

798. Richardson, H.W., and P. Gordon. "Economic and Fiscal Impacts of Metropolitan Decentralization: The Southern California Case." *Environment and Planning A* 11, 6 (1979): 643-654.

799. Rood, M.V. "Downtown Los Angeles Revisited: An Evaluation of ULI Panel Recommendations to Strengthen Retail Activity." *Urban Land* 40, 2 (1981): 12-19.

800. Roper, J.H. "Ah, Venice! An Eccentric Los Angeles Community Wants to Save Its Dream of Canals and Gondolas." *American Preservation* 2, 4 (1979): 26-38.

801. Rosenbaum, Fred. *Architects of Reform: Congregational and Community Leadership, Emanuel of San Francisco, 1849-1980.* Berkeley, CA: Western Jewish History Center, 1980.

802. Saul, E., and D. Deneri. *The San Francisco Earthquake and Fire of 1906.* Millbrae, CA: Celestial Arts, 1981.

803. Schiesl, M.J. "City Planning and the Federal Government in World War II: The Los Angeles Experience." *California History* 59, 2 (1980): 126-143.

804. Segal, H.P. "Jeff W. Hayes: Reform Boosterism and Urban Utopianism." *Oregon Historical Quarterly* 79, 4 (1978): 345-358.

805. Shumsky, N.L. "Vice Responds to Reform: San Francisco,
      1910-1914." *Journal of Urban History* 7, 1 (1980):
      31-47.

806. ———, and L.M. Springer. "San Francisco's Zone of
      Prostitution, 1880-1934." *Journal of Historic Geog-
      raphy* 7, 1 (1981): 71-94.

807. Steiner, R. *Los Angeles: The Centrifugal City*. Du-
      buque, IA: Hunt Publishing Co., 1981.

808. Stephens, G.M. "Santa Cruz Cashes in on Downtown Re-
      vitalization." *Urban Land* 38, 9 (1979): 12-17.

809. Tamari, T. "Computing in the City of Los Angeles."
      Computers, Environment and Urban Systems 7, 1/2
      (1982): 95-100.

810. Taschner, M. "Boomerang Boom: San Diego, 1941-1942."
      Journal of San Diego History 28 (Winter 1982): 1-10.

811. Tracey, C.A. "Police Function in Portland, 1867-1874,
      Pt. III." *Oregon Historical Quarterly* 80, 3 (1979):
      287-322.

812. Walker, D. *Los Angeles*. New York: St. Martin's Press,
      1981.

813. Weightman, B.A. "Arcadia in Suburbia: Orange County
      California." *Journal of Cultural Geography* 2, 1
      (1981): 55-69.

814. Whitfield, V.J. *History of Pleasant Hill, California*.
      Pleasant Hill, CA: Whitfield Books, 1981.

815. Wiersema, B., and M. Taschner. "The Selling of a City:
      Oceanside 1920-1930." *Journal of San Diego History*
      27 (Spring 1981): 71-90.

816. Wilbur, S.K. "The History of Television in Los Angeles,
      1931-1952, Pt. III." *Southern California Quarterly*
      60, 3 (1978): 255-286.

Rocky Mountain

817. Bate, K.W. "Iron City, Mormon Mining Town." *Utah
      Historical Quarterly* 50, 1 (1982): 47-58.

818. Chisum, E.D. "Boom Towns on the Union Pacific; Laramie,
      Benton, and Bear River City." *Annals of Wyoming* 53, 1
      (1981): 2-13.

819. Copper, S. "Growth Control Evolves in Boulder." *Urban*

820. Cross, N.W. "New England City, Dakota Territory: A Narrative History, 1887-1912." *North Dakota History* 47, 3 (1980): 4-10.

821. Davies, W.K. "Urban Connectivity in Montana." *Annals of Regional Science* 13, 2 (1979): 29-46.

822. Diemer, J.A. "Agriculture, Urban Development and the Public Domain in the Rocky Mountain West." *Land Economics* 55, 4 (1979): 532-536.

823. Heinerman, J. "The Mormon Meetinghouse: Reflections on Pioneer Relations and Social Life in Salt Lake City." *Utah Historical Quarterly* 50, 4 (1982): 340-353.

824. Heiss, F.W. "The Denver Regional Study: An Experiment in Reorganization." *National Civic Review* 67, 9 (1978): 407-413.

825. Larsen, L.H., and R.T. Johnson. "A Story That Never Was: North Dakota's Urban Development." *North Dakota History* 47, 4 (1980): 4-10.

826. Lewis, D.R. "La Plata, 1891-1893, Boom, Bust and Controversy." *Utah Historical Quarterly* 50, 1 (1982): 5-21.

827. Madsen, B.D. *Corinne: The Gentle Capital of Utah.* Salt Lake City, UT: Utah State Historical Society, 1980.

828. McCormick, J.S. "Red Lights in Zion: Salt Lake City's Stockade, 1908-1911." *Utah Historical Quarterly* 50, 2 (1982): 168-181.

829. Noel, T.J. *The City and the Saloon, Denver, 1856-1916.* Lincoln, NE: University of Nebraska Press, 1981.

830. Petrik, P. "Capitalists with Rooms: Prostitution in Helena, Montana, 1865-1900." *Montana* 31, 2 (1981): 28-41.

831. Remele, L. "Sewage Disposal and Local Politics at Jamestown, 1926-1929: A Case Study in North Dakota Urban History." *North Dakota History* 49, 2 (1982): 22-29.

832. Rose, M., and J.G. Clark. "Light, Heat, and Power: Energy Choices in Kansas City, Wichita, and Denver, 1900-1935." *Journal of Urban History* 5, 3 (1979): 340-364.

833. Schmidt, C.G. "An Analysis of Firm Relocation Patterns

in Metropolitan Denver, 1974-76." *The Annals of Regional Science* 13, 1 (1978): 79-91.

834. Schwantes, C.A. "Vigilantes, Grangers and the Walla Walla Outrage of June 1918." *Montana* 31, 1 (1981): 18-29.

835. Smith, C. "Public/Private Cooperation: The Fort Collins Approach." *Urban Land* 40, 6 (1981): 23-27.

836. Smith, P. *A Look at Boulder: From Settlement to City.* Boulder, CO: Pruett Publishing Co., 1981.

837. Zellick, A. "Patriots on the Rampage: Mob Action in Lewistown, 1917-1918." *Montana* 31, 1 (1981): 30-43.

Southwest

838. Baxter, J.O. "Salvador Armijo: Citizen of Albuquerque, 1823-1879." *New Mexico Historical Review* 53, 3 (1978): 219-238.

839. Biebel, C.D. "Cultural Change on the Southwest Frontier: Albuquerque Schooling, 1870-1895." *New Mexico Historical Review* 55, 3 (1980): 209-230.

840. Bissell, H.H., and B.T. Cima. "Dallas Freeway Corridor Study." *Public Roads* 45, 3 (1981): 89-94.

841. Brownell, B.A. "If You Have Seen One, You Haven't Seen Them All: Recent Trends in Southern Urban History." *Houston Review* 1 (Fall 1979): 63-80.

842. Bufkin, D. "From Mud Village to Modern Metropolis: The Urbanization of Tucson." *Journal of Arizona History* 22, 1 (1981): 63-98.

843. Bullard, R.D. *Housing Allowance in the Seventies: An Assessment of Houston's Housing Assistance Payment Program.* Houston, TX: Texas Southern University Press, 1977.

844. Cable, C. "The Architecture of Houston, Texas 1955-1977." *Council of Planning Librarians Bibliographies*, A-2 (June 1978).

845. Christian, G.L. "Rio Grande City: Prelude to the Brownsville Raid." *West Texas Historical Association Yearbook* 57 (1981): 118-132.

846. Ducker, J.H. "Workers, Townsmen, and the Governor: The Santa Fe's Engineers' Strike, 1878." *Kansas History* 5, 1 (1982): 23-32.

847. Fisher, R. "Community Organizing in Historical Perspective: A Typology." *Houston Review* 4 (Summer 1982): 75-88.

848. Garner, J.S. "The Saga of a RR Town: Calvert, Texas (1868-1918)." *Southwestern Historical Quarterly* 85, 7 (1981): 139-160.

849. Gober, P. "Shrinking Household Size and Its Effect on Urban Population Density Patterns: A Case Study of Phoenix, Arizona." *Professional Geographer* 32, 1 (1980): 55-62.

850. Grant, H.R. "Interurbans Are the Wave of the Future: Electric Railway Promotion in Texas." *Southwestern Historical Quarterly* 84, 1 (1980): 29-48.

851. Harries, K.D., and S.J. Stadler. "Determinism Revisited: Assault and Heat Stress in Dallas, 1980." *Environment and Behavior* 15, 2 (1983): 235-255.

852. Hermann, R. *Virginia City, Nevada Revisited.* Sparks, NV: Falcon Hills Press, 1981.

853. Kaplan, B.J. "Urban Development, Economic Growth, and Personal Liberty: The Rhetoric of the Houston Anti-Zoning Movements, 1947-1962." *Southwestern Historical Quarterly* 84, 2 (1980): 133-168.

854. Konig, M. "Phoenix in the 1950's: Urban Growth in the Sunbelt." *Arizona and the West* 24, 1 (1982): 19-38.

855. Lazarou, K.E. "A History of the Port of Galveston: A Constitutional Legal Overview." *Houston Review* 2 (Summer 1980): 83-95.

856. Love, P. *From Brothel to Boom Town: Yuma's Naughty Past.* Colorado Springs, CO: Little London Press, 1981.

857. Luckingham, B. "Urban Development in Arizona: The Rise of Phoenix." *Journal of Arizona History* 22, 2 (1981): 197-234.

858. ———. *The Urban Southwest: A Profile History of Albuquerque, El Paso, Phoenix, and Tucson.* El Paso, TX: Texas Western Press, 1982.

859. McClendon, B.W. "Reforming Zoning Regulations to Encourage Economic Development: Beaumont, Texas." *Urban Land* 40, 4 (1981): 3-7.

860. Melosi, M.V. *Garbage in the Cities: Refuse, Reform, and the Environment, 1880-1980.* College Station, TX: Texas A and M University Press, 1981.

861. Melzer, R. "A Death in Dawson: The Demise of a South-
      western Company Town." *New Mexico Historical Review*
      55, 4 (1980): 309-330.

862. Mihelich, D.N. "World War II and the Transformation
      of the Omaha Urban League." *Nebraska History* 60, 3
      (1979): 401-423.

863. Miller, M.V. *Economic Growth and Change along the US-
      Mexican Border*. Austin, TX: University of Texas Press,
      1982.

864. Moehring, E.P. "Public Works and the New Deal in Las
      Vegas, 1933-1940." *Nevada Historical Society Quarter-
      ly* 24, 2 (1981): 107-129.

865. O'Rourke, M.K. "Composition and Distribution of Urban
      Vegetation in the Tucson Basin." *Journal of Arid En-
      vironments* 5, 3 (1982): 235-248.

866. Simmons, M. "Governor Cuervo and the Beginnings of Al-
      buquerque: Another Look." *New Mexico Historical Re-
      view* 55, 3 (1980): 188-207.

867. Smith, B.A., and R. Ohsfeldt. "Housing Price Inflation
      in Houston: 1970-1976." *Policy Studies Journal* 8, 2
      (1979): 257-276.

868. Speer, J.B. "Pestilence and Progress: Health Reform
      in Galveston and Houston during the Nineteenth Cen-
      tury." *Houston Review* 2 (Fall 1980): 120-132.

869. Szasz, M.C. "Albuquerque Congregationalists and South-
      western Social Reform, 1900-1917." *New Mexico His-
      torical Review* 55, 3 (1980): 231-252.

870. "Texas Labor in the Thirties: Gilbert Mers and the
      Corpus Christi Waterfront Strikes." *Houston Review*
      3 (Fall 1981): 308-320.

871. Thompson, T.R. "The Devil Wagon Comes to Omaha: The
      First Decade of the Automobile, 1900-1910." *Nebraska
      History* 61, 2 (1980): 172-191.

872. Timmons, W.H. "The El Paso Area in the Mexican Period,
      1821-1848." *Southwestern Historical Quarterly* 84, 1
      (1980): 1-28.

873. Todd, C. "Metal Mining and Its Associated Industries
      in Tucson." *Journal of Arizona History* 22, 1 (1981):
      99-128.

874. Watson, C. "Brownsville: Working To Make a Texas City
      Aware of Its Heritage." *American Preservation* 2, 5
      (1979): 7-21.

875. Wehmerer, E. "The Salt River Project (SRP), Central Arizona and the Area of Greater Phoenix: A Case Study in Urbanization and Trends of Water Consumption." *Geoforum* 11, 2 (1980): 107-121.

876. Zarbin, E. "The Whole Was Done So Quietly: The Phoenix Lynchings of 1879." *Journal of Arizona History* 21, 4 (1980): 353-362.

AN AMBIENCE OF ETHNICITY

Assimilation into the American way of life has occurred
for the majority of new arrivals to the nation within the
urban environment. This common experience of ethnic groups
in urban America is well documented by Robert F. Harney and
Harold M. Troper in *Immigrants: A Portrait of the Urban Experience 1890-1930* (1975). Kathleen Conzen's "Immigrants,
Immigrant Neighborhoods, and Ethnic Identity: Historical
Issues" (*Journal of American History* 66, 3 (1979): 603-615),
is a more recent examination of the relationship between ethnic
groups and the urban environment.

Ethnicity in the American urban setting has been the focus
of numerous works in the last decade. New York, Chicago,
Pittsburgh, Boston, and Philadelphia have been carefully examined to yield information about ethnic groups within their
political boundaries. James Borchert's *Alley Life in Washington: Family, Community, Religion, and Folklife in the City
1850-1970* (1980) is an interesting study of Black family life
in the nation's capital. This work challenges the traditional
view that the urban environment broke down the social institutions that Black migrants brought into the city. Borchert
maintains that the Black migrants to Washington adapted the
patterns of their folk existence and social organization to
the conditions confronting them in the urban environment. A
new institution, the alley, was slowly incorporated into the
Black migrant's usual social organization, according to
Borchert, so that a continuation of folk experiences was
possible through use of the city's alleyways as a source of
protection and support.

New York City is the location of Deborah D. Moore's *At
Home in America: Second Generation New York Jews* (1981). Assimilation into American society for Jewish families often
means the possible loss of religious identity. Between 1920
and 1950, Moore finds that second-generation Jewish families
moved from the Lower East Side of New York City into the Bronx
and Brooklyn. In these new locations, a Jewish identity was
preserved. Jewish middle-class life was offered as an alternative to assimilation into American culture. A similar study
conducted by Carolyn and Gordon Kirk, "The Impact of the City

on Home Ownership: A Comparison of Immigrant and Native
Whites at the Turn of the Century" (*Journal of Urban History*
7, 4 (1981): 471-498), examines the ease of assimilation for
immigrant groups in face of a hostile urban environment. The
Kirks insist that only by viewing the interaction of the urban
structure and the attitudes and values of immigrant groups
can generalizations about immigrant behavior be made. Their
study of the rate of home ownership among native Whites and
immigrants indicates that the urban environment discriminated
against the immigrant groups so that home ownership was more
a possibility for native Whites than immigrants. Thus, the
urban environment is a significant factor in the ease of assimilation into American society by immigrants, according to
the Kirks.

The following list of books and journal articles is
representative of those recently written on the experience
of various ethnic groups in urban America.

General

877. Agocs, C. "Ethnic Settlement in a Metropolitan Area:
      A Typology of Communities." *Ethnicity* 8, 2 (June
      1981): 127-148.

878. Allen, I. "Suburban Preferences of Urban Ethnic Groups:
      Eight White Groups in Three Connecticut Cities, 1966."
      Social Science Quarterly 62 (March 1981): 99-107.

879. Allswareg, J.M. *The Political Behavior of Chicago's
      Ethnic Groups*. New York: Arno Press, 1980. Reprint.

880. Baylor, R.H. *Neighbors in Conflict: The Irish, Germans,
      and Italians of New York City*. Baltimore, MD:
      Johns Hopkins University Press, 1980.

881. Baylor, R.N. "Ethnic Residential Patterns in Atlanta,
      1880-1940." *Georgia Historical Quarterly* 63, 4
      (Winter 1979): 435-446.

882. Berrol, S.C. *Immigrants at School, New York City 1898-
      1914*. New York: Arno Press, 1979. Reprint.

883. Berry, B.J. "Ghetto Expansion and Single Family Housing
      Prices: Chicago, 1968-1972." *Journal of Urban Eco-
      nomics* 3, 3 (October 1976): 397-423.

884. Bodnar, J.E. *Immigration and Industrialization: Eth-
      nicity in an American Mill Town 1870-1970*. Pitts-
      burgh, PA: Pittsburgh University Press, 1977.

885. Browning, R.P., et al. "Minorities and Urban Electoral
      Change: A Longitudinal Study." *Urban Affairs Quar-
      terly* 15, 2 (December 1979): 206-228.

886. Bullard, Robert D. "Does Section 8 Promote an Ethnic
      and Economic Mix?" *Journal of Housing* 35, 2 (1978):
      364-365.

887. Capeci, D.J. "Fiorello La Guardia and the Harlem Crime
      Wave of 1940." *New-York Historical Society Quarterly*
      64, 1 (January 1980): 7-30.

888. Chudacoff, H.P. "A New Look at Ethnic Neighborhoods:
      Residential Dispersion and the Concept of Visibility
      in a Medium-Sized City." *Journal of American History*
      60, 1 (June 1973): 76-93.

889. Clark, D. "Ethnic Enterprise and Urban Development."
      Ethnicity 5, 2 (June 1978): 108-118.

890. Colburn, D.R., and G.E. Pozzetta. *America and the New
      Ethnicity*. Port Washington, NY: Kennikat Press, 1978.

891. Conzen, K.N. *Immigrant Milwaukee, 1836-1860*. Cam-
      bridge, MA: Howard University Press, 1976.

892. ————. "Immigrants, Immigrant Neighborhoods and
      Ethnic Identity: Historical Issues." *Journal of
      American History* 66, 3 (December 1979): 603-615.

893. Cotter, J.V., and L.L. Patrick. "Disease and Ethnicity
      in an Urban Environment." *Association of American
      Geographers Annual* 71, 1 (March 1981): 40-49.

894. Cummings, S. *Self-Help in Urban America: Patterns of
      Minority Economic Development*. Port Washington, NY:
      Kennikat Press, 1979.

895. D'A Jones, P., and M.G. Holli. *Ethnic Chicago*. Grand
      Rapids, MI: Wm. B. Eerdmans Publishing Co., 1980.

896. Davis, A.F., and M.H. Haller. *The Peoples of Philadel-
      phia: A History of Ethnic Groups and Lower Class Life,
      1790-1940*. Philadelphia, PA: Temple University Press,
      1973.

897. Dolan, J.P. "Immigrants in the City: New York's Irish
      and German Catholics." *Church History* 41, 3 (September
      1972): 354-368.

898. Eckert, W.A. "The Applicability of the Ethos Theory
      to Specific Ethnic Groups and the Prediction of Urban
      Political Forms." *Urban Affairs Quarterly* 11, 3
      (March 1976): 357-386.

899. Eisinger, P.K. "Ethnic Political Transition in Boston, 1884-1933, Some Lessons for Contemporary Cities." *Political Science Quarterly* 93, 2 (Summer 1978): 217-239.

900. Ernst, R. *Immigrant Life in New York City 1825-1863.* New York: Octagon Books, 1979.

901. Ford, L., and E. Griffin. "The Ghettoization of Paradise (San Diego)." *Geographical Review* 69, 2 (April 1979): 140-158.

902. Gabriel, R.A. *The Irish and Italians: Ethnics in City and Suburb.* New York: Arno Press, 1980.

903. Garonski, J. "The Racial and Ethnic Make-Up of Baltimore Neighborhoods 1850-1870." *Maryland Historical Magazine* 71, 3 (Fall 1976): 392-402.

904. Glasco, L.A. *Ethnicity and Social Structure: Irish, Germans, and Native Born of Buffalo, N.Y., 1850-1860.* New York: Arno Press, 1980. Reprint.

905. Glassberg, E. "Work, Wages and the Cost of Living: Ethnic Differences and the Poverty Line, Philadelphia 1880." *Pennsylvania History* 46, 1 (January 1979): 17-58.

906. Golant, S.M., and C.W. Jacobsen. "Factors Underlying the Decentralized Residential Location of Chicago's Ethnic Population." *Ethnicity* 5 (December 1978): 379-397.

907. Greeley, A.M. "Ethnic Domestic Architecture in Chicago." *Ethnicity* 6 (June 1979): 137-146.

908. Guest, A.M. "The Changing Racial Composition of Suburbs 1950-1970." *Urban Affairs Quarterly* 14, 2 (December 1978): 195-206.

909. ————. "Suburbanization of Ethnic Groups." *Sociology and Social Research; an International Journal* 64 (July 1980): 497-513.

910. Hamilton, C.V. "Patron-Recipient Relationship and Minority Politics in New York City." *Political Science Quarterly* 94, 2 (Summer 1979): 11-27.

911. Hardy, M.A. "Occupational Mobility and Nativity-Ethnicity in Indianapolis, 1850-1860." *Social Forces* 57, 1 (September 1978): 205-221.

912. Harney, R., and H. Troper. *Immigrants: A Portrait of the Urban Experience 1890-1930.* Toronto: Van Nostrand and Reinhold, 1975.

913. Haynes, R.V. "The Houston Mutiny and Riot of 1917."
      Southwestern Historical Quarterly 76, 4 (April 1973):
      418-439.

914. Hecker, M. *Ethnic America, 1970-1977.* New York: Oceana
      Publications, 1979.

915. Hennessey, M.M. "Race and Violence in Reconstruction
      New Orleans: 1868 Riot." *Louisiana History* 20, 1
      (Winter 1979): 77-92.

916. Hill, R.F. *Exploring the Dimensions of Ethnicity: A
      Study of Status, Culture, and Identity.* New York:
      Arno Press, 1980. Reprint.

917. Holli, M.G., and F.A. Jones. *The Ethnic Frontier: Es-
      says in the History of Group Survival in Chicago and
      the Midwest.* Grand Rapids, MI: Wm. B. Eerdmans Pub-
      lishing Co., 1977.

918. Hurst, M. "Integration, Freedom of Choice and Community
      Control in Nineteenth-Century Brooklyn." *Journal of
      Ethnic Studies* 3, 3 (Fall 1975): 33-56.

919. *Immigrants and Religion in Urban America.* Philadelphia,
      PA: Temple University Press, 1977.

920. Janis, R. "Ethnic Mixture and the Presence of Cultural
      Pluralism in the Church Communities of Detroit 1880-
      1940." *Mid-America* 61, 2 (April/July 1979): 99-116.

921. ————. "Flirtation and Flight: Alternatives to Ethnic
      Confrontation in White Anglo-American Protestant De-
      troit 1880-1940." *Journal of Ethnic Studies* 6, 2
      (Summer 1978): 1-18.

922. Jensen, R. "Cities Re-Elect Roosevelt: Ethnicity, Re-
      ligion, and Class in 1940." *Ethnicity* 8, 2 (June 1981):
      189-195.

923. Jones, P., and M.G. Holli. *Ethnic Chicago.* Grand
      Rapids, MI: Wm. B. Eerdmans Publishing Co., 1981.

924. Kantrowitz, N. "Racial and Ethnic Residential Segrega-
      tion in Boston, 1830-1970." *American Academy of
      Political and Social Science Annals* 441 (January 1979):
      41-54.

925. Karnig, A.K., and S. Welch. "Sex and Ethnic Differences
      in Municipal Representation." *Social Science Quar-
      terly* 60, 4 (December 1979): 465-481.

926. Kessner, Thomas. "Jobs, Ghettos and the Urban Economy,
      1880-1935." *American Jewish History* 71 (December
      1981): 218-238.

927. Kirk, C.T., and G.W. Kirk. "The Impact of the City on
      Home Ownership: A Comparison of Immigrants and Native
      Whites at the Turn of the Century." *Journal of Urban
      History* 7 (August 1981): 471-498.

928. Kirk, G.W. "Migration, Mobility and the Transformation
      of the Social Structure in an Immigrant Country:
      Holland, Michigan 1850-1880." *Journal of Social History* 7, 2 (Winter 1974): 142-164.

929. Kranse, C.A. "Urbanization Without Breakdown: Italian,
      Jewish and Slavic Immigrant Women in Pittsburgh 1906-
      1945." *Journal of Urban History* 4, 3 (May 1978): 291-
      306.

930. Leinenweber, C. "The Class and Ethnic Bases of New York
      City Socialism, 1904-1915." *Labor History* 22, 1
      (Winter 1981): 31-56.

931. Leonard, H.B. "The Immigrants Protective League of
      Chicago, 1908-1921." *Journal of Illinois State Historical Society* 66, 3 (Autumn 1973): 271-284.

932. Levine, D.A. *Internal Combustion: The Races in Detroit
      1915-1926.* Westport, CT: Greenwood Press, 1976.

933. Lopez, M.M. "Patterns of Interethnic Residential Segre-
      gation in the Urban Southwest, 1960 and 1970." *Social
      Science Quarterly* 62, 1 (March 1981): 50-63.

934. Massey, D. "Social Class and Ethnic Segregation: Re-
      consideration of Methods and Conclusions." *American
      Sociological Review* 46, 3 (October 1981): 641-650.

935. Merriam, P.G. "The Other Portland: A Statistical Note
      on Foreign-Born, 1860-1910." *Oregon Historical
      Quarterly* 80, 3 (1979): 258-268.

936. Meyers, M. "Urban Ghetto." *Crisis* 88 (January/February
      1981): 42-46.

937. Miller, R.M., and T.D. Marzik. *Immigrants and Religion
      in Urban America.* Philadelphia, PA: Temple University
      Press, 1977.

938. Morgan, M., and H.H. Golden. "Immigrant Families in
      an Industrial City: A Study of Households in Holyoke,
      1880." *Journal of Family History* 4, 1 (Spring 1979):
      59-68.

939. Pap, M. *Ethnic Communities of Cleveland: A Reference
      Work.* Cleveland, OH: John Carroll University Press,
      1973.

940. Rabinowitz, H.N. *Race Relations in the Urban South 1865-1890*. New York: Oxford University Press, 1978.

941. Shaw, D.V. *The Making of an Immigrant City: Ethnic and Cultural Conflict in Jersey City, N.J. 1850-1877*. New York: Arno Press, 1976. Reprint.

942. Shover, J.L. "Ethnicity and Religion in Philadelphia Politics 1924-40." *American Quarterly* 25, 5 (December 1973): 499-515.

943. Spengler, P.A. *Yankee, Swedish and Italian Acculturation and Economic Mobility in Jamestown, New York from 1860-1920*. New York: Arno Press, 1980. Reprint.

944. Stack, J.F. *International Conflict in an American City: Boston's Irish, Italians and Jews 1935-1944*. Westport, CT: Greenwood Press, 1979.

945. Stoloff, D.L. "Minority Ethnic Television in Los Angeles: A Multi-Cultural Alternative." *Urban Review* 13, 3 (1981): 147-160.

946. Thernstrom, S., ed. *Harvard Encyclopedia of American Ethnic Groups*. Cambridge, MA: Belknap Press of Harvard University Press, 1980.

947. Thomas, J.C. "Budget-Cutting and Minority Employment in City Governments: Lessons from Cincinnati." *Public Personnel Management* 7, 2 (May/June 1978): 155-161.

948. Ward, D. "The Ethnic Ghetto in the U.S.: Past and Present." *Transactions. Institute of British Geographers* 7, 3 (1982): 257-275.

949. Watts, A.D., and T.M. Watts. "Minorities and Urban Crime: Are They the Cause or the Victims?" *Urban Affairs Quarterly* 16, 4 (June 1981): 423-436.

950. Weber, M.P. "Residential and Occupational Patterns of Ethnic Minorities in Nineteenth Century Pittsburgh." *Pennsylvania History* 44, 4 (October 1977): 317-334.

951. White, M.J. "Job Suburbanization, Zoning and the Welfare of Urban Minority Groups." *Journal of Urban Economics* 5, 2 (April 1978): 219-240.

952. Zunz, O. "The Organization of the American City in the Late Nineteenth Century: Ethnic Structural and Spatial Arrangement in Detroit." *Journal of Urban History* 3 (August 1977): 443-466.

Asian

953. Armentrout-Ma, E. "Urban Chinese at the Sinitic Fron-
     tier: Social Organizations in United States' China-
     towns 1849-1898." *Modern Asian Studies* 17, 1 (1983):
     67-136.

954. Blackburn, G.M., and S.L. Richards. "The Chinese of
     Virginia City, Nevada, 1870." *Amerasia Journal* 7, 1
     (Spring 1980): 51-72.

955. Desbarats, J. "Thai Migration to Los Angeles." *Geo-
     graphical Review* 69, 3 (July 1979): 302-318.

956. Dicker, L.M. *The Chinese in San Francisco: A Pictorial
     History.* New York: Dover, 1980.

957. Dong, L., and M.K. Ham. "Chinatown Chinese: The San
     Francisco Dialect." *Amerasia Journal* 7, 1 (1980):
     1-30.

958. Dunnigan, T. "Segmentary Kinship in an Urban Society:
     The Hmong of St. Paul-Minneapolis." *Anthropological
     Quarterly* 55, 3 (1982): 126-134.

959. Fisher, M.P. "Creative and Ethnic Identity: Asian In-
     dians in the New York City Area." *Urban Anthropology*
     7, 3 (Fall 1978): 271-286.

960. Glenn, E.N. "Occupational Ghettoization: Japanese-
     American Women and Domestic Service, 1905-1970."
     Ethnicity 8, 4 (December 1981): 352-386.

961. Haines, D., et al. "Case for Exploratory Fieldwork:
     Understanding the Adjustment of Vietnamese Refugees
     in the Washington Area." *Anthropological Quarterly*
     54, 2 (April 1981): 94-102.

962. Hamill, P., and M. Miyamoto. "The Japanning of New
     York." *New York Magazine* 14 (August 17, 1981): 16-23.

963. Hata, D.T. *Undesirable: Early Immigrants and the Anti-
     Japanese Movement in San Francisco 1892-1893, Prelude
     to Exclusion.* New York: Arno Press, 1978. Reprint.

964. Hemminger, C. "Little Manila: The Filipino in Stockton
     Prior to World War II." *Pacific Historian* 24 (Spring
     1980): 21-34; 207-220.

965. Johnson, L.D. "Equal Rights and the Heathen Chinese:
     Black Activism in San Francisco." *Western Historical
     Quarterly* 11, 1 (1980): 57-68.

966. Kim, B.C. "Problems and Service Needs of Asian-Americans

in Chicago: An Empirical Study." *Amerasia Journal* 5 (1978): 23-44.

967. Kim, I. *New Urban Immigrants: The Korean Community in New York*. Princeton, NJ: Princeton University Press, 1981.

968. Kwong, P. *Chinatown, New York: Labor and Politics 1930-1950*. New York: Monthly Review Press, 1979.

969. Li, P.S. "Occupational Achievements and Kinship Assistance among Chinese Immigrants in Chicago." *Sociological Quarterly* 18 (Autumn 1977): 478-489.

970. Light, I. "From Vice District to Tourist Attraction: The Moral Career of American Chinatowns 1880-1940." *Pacific Historical Review* 43, 3 (August 1974): 367-394.

971. Lyman, S.M. "Conflict and the Web of Group Affiliation in San Francisco's Chinatown, 1850-1910." *Pacific Historical Review* 43, 3 (August 1974): 473-499.

972. Magnaghi, R.M. "Virginia City's Chinese Community, 1860-1880." *Nevada Historical Society Quarterly* 24 (Summer 1981): 130-157.

973. Modell, J. *The Economics and Politics of Racial Accommodation: The Japanese of Los Angeles 1900-1942*. Urbana, IL: University of Illinois Press, 1977.

974. Schwartz, H. "A Union Combats Racism: The ILWU's Japanese-American Stockton Incident of 1945." *Southern California Quarterly* 62 (Summer 1980): 161-176.

975. Scott, G.M. "The Hmong Refugee Community in San Diego: Theoretical and Practical Implications of Its Continuing Ethnic Solidarity." *Anthropological Quarterly* 55, 3 (1982): 146-160.

976. Starr, P., and A. Roberts. "Attitudes toward Indochinese Refugees: An Empirical Study." *Journal of Refugee Resettlement* 1 (August 1981): 51-61.

977. Yong, P. "Chinese Labor in Early San Francisco: Racial Segmentation and Industrial Expansion." *Amerasia Journal* 8 (1981): 69-92.

978. Yu, C.Y. "A History of San Francisco Chinatown Housing." *Amerasia Journal* 8 (Spring/Summer 1981): 93-110.

979. Entry deleted.

Blacks

980. Anderson, E.F. *The Development of Leadership and Or-*
      ganization Building in the Black Community of Los
      Angeles from 1900 through World War II. Saratoga,
      CA: Century Twenty-One Publications, 1980.

981. Aschenbreuner, J. *Lifelines: Black Families in Chicago.*
      New York: Holt, Rinehart and Winston, 1975.

982. Bent, Devin. "Partisan Elections and Public Policy: Re-
      sponse to Black Demands in Large American Cities."
      Journal of Black Studies 12, 3 (March 1982): 291-314.

983. Bergesen, A. "Race Riots of 1967: An Analysis of Police
      Violence in Detroit and Newark." *Journal of Black*
      Studies 12 (1982): 261-274.

984. Berry, B.J. *The Open Housing Question: Race and Housing*
      in Chicago, 1966-1976. Cambridge, MA: Ballinger
      Publishing Co., 1979.

985. Betten, N., and R.A. Mohl. "The Evolution of Racism in
      an Industrial City 1906-1940: A Case Study of Gary,
      Indiana." *Journal of Negro History* 59, 1 (January
      1974): 51-64.

986. Blassingame, J.W. "Before the Ghetto: The Making of
      the Black Community in Savannah, Georgia 1865-1880."
      Journal of Social History 6, 4 (Summer 1973): 463-
      488.

987. ———. *Black New Orleans 1860-1880.* Chicago, IL:
      University of Chicago Press, 1973.

988. Bonnett, A.W. "Structures Adaptation of Black Migrants
      from the Caribbean: An Examination of an Indigenous
      Banking System in Brooklyn." *Phylon* 42, 4 (December
      1981): 346-355.

989. Bragow, D.H. "Status of Negroes in a Southern Port City
      in the Progressive Era: Pensacola 1896-1920." *Florida*
      Historical Quarterly 51, 3 (January 1973): 281-302.

990. Branton, W.A. "Little Rock Revisited: Desegregation
      and Resegregation." *Journal of Negro Education* 52
      (Summer 1983): 250-269.

991. Broussard, A.S. "Organizing the Black Community in the
      San Francisco Bay Area, 1915-1930." *Arizona West* 23
      (Winter 1981): 335-354.

992. Bulmer, M., and C.S. Johnson. "Robert E. Park and the
      Research Methods of the Chicago Commission on Race

Relations 1919-1922: An Early Experiment in Applied
Social Research." *Ethnic and Racial Studies* 4 (July
1981): 289-306.

993. Button, J., and R. Scher. "Impact of the Civil Rights
Movement: Perceptions of Black Municipal Service
Changes." *Social Science Quarterly* 60, 3 (1979):
497-510.

994. Christian, C.M., and S.J. Bennett. "The Reclamation
of Industrial Building Vacancies: Chicago's Black
Community." *Urban Affairs Quarterly* 13 (September
1977): 109-116.

995. Clark, T.A. *Blacks in Suburbs, a National Perspective.*
Piscataway, NJ: Center for Urban Policy Research,
1979.

996. Clay, P.L. "The Process of Black Suburbanization."
Urban Affairs Quarterly 14, 4 (June 1979): 405-424.

997. Collins, K.E. *Black Los Angeles: The Maturing of the
Ghetto 1940-1950.* Saratoga, CA: Century Twenty-One
Publications, 1980.

998. Connolly, H.X. *A Ghetto Grows in Brooklyn.* New York:
New York University Press, 1977.

999. Curry, J.P., and G.D. Scriven. "The Relationship be-
tween Apartment Living and Fertility for Blacks,
Mexican-Americans, and other Americans in Racine,
Wisconsin." *Demography* 15 (November 1978): 477-486.

1000. Curry, L.P. *The Free Black in Urban America, 1800-1850:
The Shadow of the Dream.* Chicago, IL: University of
Chicago Press, 1981.

1001. Daniels, D.H. "Black Americans in Early San Francisco."
Center Magazine 13 (May 1980): 49-63.

1002. ————. *Pioneer Urbanites: A Social and Cultural History of Black San Francisco.* Philadelphia, PA:
Temple University Press, 1980.

1003. Darden, J.T. "The Residential Segregation of Blacks
in Detroit, 1960-1970." *International Journal of
Comparative Sociology* 17 (March/June 1976): 84-91.

1004. ————. "Residential Segregation of Blacks in the
Suburbs: The Michigan Example." *Geographical Survey*
5 (July 1976): 7-16.

1005. Davis, D., and E. Casetti. "Do Black Students Wish to
Live in Integrated, Socially Homogeneous Neighborhoods?" *Ekistics* 46, 275 (1979): 110-114.

1006. Day, J., and H.J. Kedro. "Free Blacks in St. Louis:
       Antebellum Condition, Emancipation and the Postwar
       Era." *Bulletin Missouri Historical Society* 30, 2
       (January 1974): 117-135.

1007. Denowitz, R.M. "Racial Succession in New York City,
       1960-70." *Social Forces* 59, 2 (December 1980): 440-
       455.

1008. Eisinger, P.K. "Black Employment in Municipal Jobs:
       The Impact of Black Political Power." *American
       Political Science Review* 76, 2 (June 1982): 380-392.

1009. Ellis, A.L. *The Black Power Brokers*. Saratoga, CA:
       Century Twenty-One, 1980.

1010. Ellsworth, S. *Death in a Promised Land: The Tulsa
       Race Riot of 1921*. Baton Rouge, LA: Louisiana State
       University Press, 1981.

1011. Engstrom, R.L., and M.D. McDonald. "The Election of
       Blacks to City Councils: Clarifying the Impact of
       Electoral Arrangements on the Seats/Population Re-
       lationship." *American Political Science Review* 75, 2
       (June 1981): 344-354.

1012. Erickson, R.A., and T.K. Miller. "Race and Resources
       in Large American Cities: An Examination of Intra-
       urban and Interregional Variations." *Urban Affairs
       Quarterly* 13, 4 (1978): 401-420.

1013. Franklin, V.P. *The Education of Black Philadelphia:
       The Social and Educational History of a Minority
       Community 1900-1950*. Philadelphia, PA: University of
       Pennsylvania Press, 1979.

1014. Frey, William H. "Black In-Migration, White Flight,
       and the Changing Economic Base of the Central City."
       American Journal of Sociology 85, 6 (May 1980): 1396-
       1447.

1015. ————. "Central City White Flight." *American Socio-
       logical Review* 44, 3 (1979): 425-448.

1016. Fuerst, J.S. "Report Card: Chicago's All-Black Schools."
       Public Interest 64 (Summer 1981): 79-91.

1017. ————, and R. Petty. "Black Housing in Chicago."
       Public Interest 52 (Summer 1979): 103-110.

1018. Fujita, K. *Black Workers' Struggles in Detroit's Auto
       Industry, 1935-1975*. Saratoga, CA: Century Twenty-
       One Publications, 1980.

1019. Furstenberg, F. "The Origins of the Female-Headed
      Black Family: The Impact of the Urban Experience."
      Journal of Interdisciplinary History 6, 2 (Autumn
      1975): 211-234.

1020. Garofalo, C. "Black-White Occupational Distribution
      in Miami during World War I." *Prologue* 5, 2 (Summer
      1973): 81-99.

1021. Goldin, C.D. *Urban Slavery in the American South
      1820-1860: A Quantitative History.* Chicago, IL:
      University of Chicago Press, 1976.

1022. Goldstein, M. "Preface to the Rise of Booker T. Wash-
      ington: A View from New York City of the Demise of
      Independent Black Politics." *Journal of Negro His-
      tory* 62 (1977).

1023. Goldstein, M.L. "Black Power and the Rise of Bureau-
      cratic Autonomy in New York City Politics: The Case
      of Harlem Hospital, 1912-1931." *Phylon* 41, 2 (1980):
      187-201.

1024. Goodfriend, J.D. "Burghers and Blacks: The Evolution
      of a Slave Society at New Amsterdam." *New York His-
      tory* 59, 2 (April 1978): 125-144.

1025. Green, R.D. "Urban Industry, Black Resistance, and
      Racial Restriction in the Antebellum South: A General
      Model and Case Study in Urban Virginia." *Journal of
      Economic History* 41, 1 (March 1981): 189-191; 199-
      201.

1026. Greenwood, M. "A Simultaneous-Equations Model of White
      and Non-White Migration and Urban Change." *Economic
      Inquiry* 14 (March 1976): 1-16.

1027. Guest, A.M. "The Changing Racial Composition of
      Suburbs 1950-1970." *Urban Affairs Quarterly* 14, 2
      (December 1978): 195-206.

1028. Harris, C.V. "Reforms in Government Control of Negroes
      in Birmingham, Alabama, 1890-1920." *Journal of
      Southern History* 38, 4 (November 1972): 567-600.

1029. Harris, W. "Work and the Family in Black Atlanta,
      1880." *Journal of Social History* 9, 3 (Spring 1976):
      319-330.

1030. Hart, G.H.T. "Residential Segregation and Neighbor-
      hood Change in Long Beach, California." *South
      African Geographical Journal* 62, 2 (1980): 121-134.

1031. Henderson, L.J. *Administrative Advocacy: Black Ad-
      ministrators in Urban Bureaucracy.* Palo Alto, CA:
      R and E Research Association, 1979.

1032. Herschberg, T., et al. "Tale of Three Cities: Blacks
      and Immigrants in Philadelphia 1850-1880, 1930-1970."
      *American Academy of Political and Social Science
      Annals,* 441 (January 1979): 55-81.

1033. Howard, J.R., and R.C. Smith. "Urban Black Politics."
      *American Academy of Political and Social Science
      Annals* 439 (September 1978): 1-150.

1034. Hyclak, T. "Note on the Relative Earnings of Central
      City Blacks." *Journal of Human Resources* 16, 2
      (Spring 1981): 304-313.

1035. Jackson, P. "Black Charity in Progressive Era Chi-
      cago." *Social Service Review* 78 (September 1978):
      400-427.

1036. Johnsen, L.D. "Equal Rights and the 'Heathen Chinee':
      Black Activism in San Francisco 1865-1875." *Western
      Historical Quarterly* 11, 1 (January 1980): 67-68.

1037. Jones, F.C. "Black Americans and the City: A His-
      torical Survey." *Journal of Negro History* 58, 1
      (January 1973): 261-282.

1038. Jones, M.E. *Black Migration in the U.S. with Emphasis
      on Selected Central Cities.* Saratoga, CA: Century
      Twenty-One Publications, 1980.

1039. Kaplan, B.J. "Race, Income, and Ethnicity: Residential
      Change in a Houston Community, 1920-1970." *Houston
      Review* 3 (Winter 1981): 178-202.

1040. Kapsis, R.E. "Black Ghetto Diversity and Anomie: A
      Sociopolitical View." *American Journal of Sociology*
      83, 5 (March 1978): 1132-1153.

1041. Karnig, A.K. "Black Economic, Political, and Cultural
      Development: Does City Size Make a Difference?"
      Social Forces 57, 4 (June 1979): 1194-1211.

1042. Katzman, D.M. *Before the Ghetto: Black Detroit in the
      Nineteenth Century.* Urbana, IL: University of Illi-
      nois Press, 1975.

1043. Keller, E.J. "Electoral Politics in Gary: Mayoral
      Performance, Organization, and the Political Economy
      of the Black Vote." *Urban Affairs Quarterly* 15, 1
      (September 1979): 43-64.

1044. Entry deleted.

1045. Kellogg, J. "The Formation of Black Residential Areas in Lexington, Kentucky, 1865-1887." *Journal of Southern History* 48 (February 1982): 21-52.

1046. ———. "Negro Urban Clusters in the Post-Bellum South." *Geographical Review* 67, 3 (July 1977): 310-321.

1047. Kornweibel, T., ed. *In Search of the Promised Land: Essays in Black Urban History*. Port Washington, NY: Kennikat Press, 1981.

1048. Kusmer, K. "Black Urban History in the U.S., Retrospect and Prospect." *Black History: Trends in History* 3, 1 (1982): 71-92.

1049. Kusmer, K.L. *A Ghetto Takes Shape: Black Cleveland 1870-1930*. Urbana, IL: University of Illinois Press, 1978.

1050. LaGuerre, M.S. "Internal Dependency: The Structural Position of the Black Ghetto in American Society." *Journal of Ethnic Studies* 6, 4 (Winter 1979): 29-44.

1051. Lake, W., and S.C. Cutter. "Typology of Black Suburbanization in New Jersey since 1970." *Geographical Review* 70, 2 (April 1980): 167-181.

1052. Latimer, M.K. "Black Political Representation in Southern Cities: Election Systems and Other Casual Variables." *Urban Affairs Quarterly* 15, 1 (September 1979): 65-86.

1053. Levine, D.A. *Internal Combustion: The Races in Detroit, 1915-1926*. Westport, CT: Greenwood Press, 1976.

1054. Lieske, J.A. "Inadvertent Empirical Theory: A Critique of the J Curve Theory and the Black Urban Riots." *Political Methodology* 6, 1 (1979): 29-62.

1055. Lim, G.C. "Discrimination, Time-Lag, and Assessment Unequity in Black Neighborhoods." *Review of Black Political Economy* (Fall 1982): 15-28.

1056. Lynch, H.P. *The Black Urban Condition: A Documentary History 1866-1971*. New York: Thomas Y. Crowell, 1973.

1057. Marshall, H.H., and J.M. Stahara. "Determinants of Black Suburbanization: Regional and Suburban Size Category Patterns." *Sociological Quarterly* 20, 2 (1979): 237-254.

1058. Marineau, W.H. "Informal Social Ties among Urban
       Black Americans: Some New Data and a Review of the
       Problem." *Journal of Black Studies* 8, 1 (September
       1977): 83-194.

1059. Massa, A. "Black Women in the White City." *Journal
       of American Studies* 8 (1974): 319-339.

1060. Massey, D.S. "Effects of Socioeconomic Factors on
       the Residential Segregation of Blacks and Spanish-
       Americans in U.S. Urbanized Areas." *American Socio-
       logical Review* 44, 6 (December 1979): 1015-1022.

1061. Meier, A., and E. Rudwick. *Black Detroit and the Rise
       of the UAW.* New York: Oxford University Press,
       1979.

1062. ————. "Negro Boycotts of Segregated Streetcars in
       Virginia 1904-1907." *Virginia Magazine of History
       and Biography* 81, 4 (October 1973): 479-487.

1063. Ment, D. "Corporations, Unions, and Blacks: The
       Struggle for Power in American Industrial Cities."
       Journal of Urban History 7 (February 1981): 247-254.

1064. Moynihan, D.P. "Patterns of Ethnic Succession: Blacks
       and Hispanics in New York City." *Political Science
       Quarterly* 94, 1 (Spring 1979): 1-14.

1065. Naison, M. "Communism and Harlem Intellectuals in
       the Popular Front: Antifascism and the Politics of
       Black Culture." *Journal of Ethnic Studies* 9, 1
       (Spring 1981): 1-25.

1066. Nicholas, W.W. "Community Safety and Criminal Activity
       in Black Suburbs." *Journal of Black Studies* 9, 3
       (1979): 311-334.

1067. Nielson, D.G. *Black Ethos: Northern Urban Negro Life
       1890-1930.* Westport, CT: Greenwood Press, 1977.

1068. Perdue, R.E. *The Negro in Savannah 1865-1900.* New
       York: Exposition Press, 1973.

1069. Pleck, E.H. *Black Migration and Poverty, Boston 1865-
       1900.* New York: Academic Press, 1979.

1070. Radzialowski, T. "The Competition for Jobs and Racial
       Stereotypes: Poles and Blacks in Chicago." *Polish-
       American Studies* 33 (Autumn 1976): 5-18.

1071. Ramsey, J.G. "The Education of Black Philadelphia:
       The School and Educational History of a Minority
       Community." *Urban Education* 15, 2 (July 1980): 251-
       254.

1072. Renshaw, P. "The Black Ghetto 1890-1940." *Journal of American Studies* 8, 1 (April 1974): 41-59.

1073. Roof, W.C. "Race and Residence in American Cities." *Annals of the American Academy of Political and Social Science* (January 1979): entire issue.

1074. ————. "Southern Birth and Racial Residential Segregation: The Case of Northern Cities." *American Journal of Sociology* 86, 2 (1981): 350-358.

1075. Ryon, R.N. "An Ambiguous Legacy, Baltimore Blacks and the CIO 1936-1941." *Journal of Negro History* 65, 1 (Winter 1980): 18-33.

1076. Sheldon, M.B. "Black-White Relations in Richmond, Virginia 1792-1820." *Journal of Southern History* 45, 1 (February 1979): 27-44.

1077. Shogien, R., and T. Craig. *The Detroit Race Riot: A Study in Violence.* New York: Oxford University Press, 1977.

1078. Smith, A.N. "Blacks and the Los Angeles Municipal Transit System 1941-1945." *Urbanism Past and Present* 6, 1/2 (Winter/Spring 1981): 25-30.

1079. Snorgrass, W. "The Black Press in San Francisco Bay Area, 1865-1900." *California History* 60 (Winter 1981-82): 306-317.

1080. Snow, D.A., and P.J. Leahy. "Making of a Black Slum-Ghetto: A Case Study of Neighborhood Transition." *Journal of Applied Behavior Science* 16, 4 (1980): 459-481.

1081. Spain, D. "Race Relations and Residential Segregation in New Orleans: Centuries of Paradox." *Annals of the American Academy of Political and Social Science* 441 (January 1979): 82-96.

1082. Straszherm, M.R. "Discrimination and the Spatial Characteristics at the Urban Market of Black Workers." *Journal of Urban Economics* 7, 1 (1980): 119-140.

1083. Taylor, D.V. *Blacks in Minnesota: A Preliminary Guide to Historical Sources.* St. Paul: Minnesota Historical Society, 1976.

1084. Taylor, Q. "The Afro-American Communities of Seattle and Portland during the 1940's." *Arizona and the West* 23, 2 (Summer 1981): 109-126.

1085. Taylor, R.L. "Black Ethnicity and the Persistence of

Ethnogenesis." *American Journal of Sociology* 84, 6 (May 1979): 1401-1423.

1086.   Thomas, H.A. "Victims of Circumstances: Negroes in a Southern Town 1865-1880." *Register of the Kentucky Historical Society* 71, 3 (July 1973): 253-271.

1087.   Vrooman, J., and S. Greenfield. "Are Blacks Making It in the Suburbs? Some New Evidence on Intrametropolitan Spatial Segmentation." *Journal of Urban Economics* 7, 2 (March 1980): 159-167.

1088.   Watts, E.J. "Black and Blue: Afro-American Police Officers in Twentieth-Century St. Louis." *Journal of Urban History* 7 (February 1981): 131-168.

1089.   Weiss, N.J. *The National Urban League 1910-1940.* New York: Oxford University Press, 1974.

1090.   Wilder, M.G. *Black Assimilation in the Urban Environment: The Impact of Migration and Mobility.* Palo Alto, CA: Rand E Research Associates Inc., 1979.

1091.   Williams, L. "Newcomers to the City: A Study of Black Populism Growth in Toledo, Ohio 1910-1930." *Ohio History* 89 (Winter 1980): 5-24.

1092.   Williams, M.D. "Belmer Diverse Life Styles in a Pittsburgh Black Neighborhood." *Ethnic Groups* 3, 1 (1980): 23-54.

1093.   Wojniusz, H.K. "Racial Hostility among Blacks in Chicago." *Journal of Black Studies* 9 (March 1979): 40-59.

1094.   Wurdock, C.J. "Neighborhood Racial Transition: A Study of the Role of White Flight." *Urban Affairs Quarterly* 17, 1 (September 1981): 75-89.

1095.   Yearwood, L. "National Afro-American Organizations in Urban Communitieś." *Journal of Black Studies* 8, 4 (June 1978): 423-438.

German

1096.   Dobbert, G.A. *The Disintegration of an Immigrant Community: The Cincinnati Germans 1870-1920.* New York: Arno Press, 1980. Reprint.

1097.   Holli, G.M. "Teuton vs. Slav: The Great War Sinks Chicago's German Kultur." *Ethnicity* 8, 4 (December 1981): 406-451.

1098. Keeth, K. "Sankt Antonius: Germans in the Alamo City in the 1850's." *Southwestern Historical Quarterly* 76, 2 (October 1972): 183-202.

1099. Knoche, C.H. *The German Immigrant Press in Milwaukee.* New York: Arno Press, 1980. Reprint.

1100. Kuhm, H.W. "Ve Goink Milvowkee." *Historical Messenger* 31 (Winter 1975): 106-113.

1101. Olson, A. "The Nature of an Immigrant Community: St. Louis Germans 1850-1920." *Missouri Historical Review* 66, 3 (April 1972): 342-359.

1102. Olson, A.L. *St. Louis Germans 1850-1920: The Nature of an Immigrant Community and Its Relation to the Assimilation Process.* New York: Arno Press, 1980. Reprint.

1103. Scott, C.H. "Hoosier Kulturkampf: Anglo-German Cultural Conflicts in Fort Wayne, 1840-1920." *Journal of German-American Studies* 15, 1 (March 1980): 9-18.

Greek

1104. Buxbaum, E.C. *The Greek American Group of Tarpon Springs, Florida: A Study of Ethnic Identification and Acculturation.* New York: Arno Press, 1980. Reprint.

1105. Costantakos, C.M. *The American Greek Subculture: Processes of Continuity.* New York: Arno Press, 1980. Reprint.

1106. Daskarolis, G.P. "San Francisco's Greek Colony: Evolution of an Ethnic Community 1890-1945." *California History* 60, 2 (Summer 1981): 114-133.

1107. Ellis, A.W. "The Greek Community in Atlanta 1900-1923." *Georgia Historical Quarterly* 58, 4 (Winter 1974): 400-408.

1108. Gizelis, G. *Narrative Rhetorical Devices of Persuasion in the Greek Community of Philadelphia.* New York: Arno Press, 1980. Reprint.

1109. Simon, A.J. "Ethnicity as a Cognitive Model: Identity Variations in a Greek Immigrant Community." *Ethnic Groups* 2, 2 (1979): 133-154.

1110. Stephanides, M. "The Greek Community in Louisville." *Filson Club Historical Quarterly* 55, 1 (January 1981): 5-26.

1111. Troy-Lovell, L.A. "Ethnic Occupational Structures: Greeks in the Pizza Business." *Ethnicity* 8, 1 (March 1981): 82-95.

Hispanic

1112. Boone, M.S. "Social Structure of a Low-Density Cultural Group: Cubans in Washington, D.C." *Anthropological Quarterly* 54, 2 (April 1981): 103-109.

1113. Castillo, G. "La Familia Chicana: Social Changes in the Chicano Family of Los Angeles, 1850-1880." *Journal of Ethnic Studies* 3, 1 (Spring 1975): 41-58.

1114. Comer, J.C. "Street Level Bureaucracy and Political Support: Some Findings on Mexican Americans." *Urban Affairs Quarterly* 14, 2 (December 1978): 207-227.

1115. Deagan, K. "Spanish St. Augustine: America's First Melting Pot." *Archaeology* 33, 5 (September/October 1980): 22-30.

1116. DiMartino, D.R. "Human Services Utilization by Older Hispanics in Nebraska." *Review of Applied Urban Research* 9 (January 1981): 1-8.

1117. "From Barrio to City Hall; San Antonio." *Economist* 279 (April 11-17, 1981): 28.

1118. Garcia, J.R. "History of Chicanos in Chicago Heights." *Aztlan* 7, 1 (Summer 1976): 291-306.

1119. Garcia, M.T. "The Chicana in American History: The Mexican Women of El Paso, 1880-1920: A Case Study." *Pacific Historical Review* 49, 2 (May 1980): 315-338.

1120. ————. *Desert Immigrants: The Mexicans of El Paso 1880-1920*. New Haven, CT: Yale University Press, 1981.

1121. Garcia, R.A. "Class, Consciousness, and Ideology: The Mexican Community of San Antonio, Texas: 1930-1940." *Aztlan* 9, 1 (1978): 23-69.

1122. Hoffman, A. "Stimulus to Repatriation: The 1931 Federal Deportation Drive and the Los Angeles Mexican Community." *Pacific Historical Review* 42, 2 (May 1973): 205-219.

1123. Keefe, Susan E. "Urbanization, Acculturation, and Extended Family Ties: Mexican Americans in Cities." *American Ethnologist* 6, 2 (1979): 349-365.

1124. Kernstock, E.N. *How New Migrants Behave Politically: The Puerto Ricans in Hartford, 1970.* New York: Arno Press, 1981. Reprint.

1125. Kerr, L.A. "Chicano Settlements in Chicago: A Brief History." *Journal of Ethnic Studies* 2, 4 (Winter 1975): 22–32.

1126. Larralde, C. *Carlos Esparza: A Chicago Chronicle.* San Francisco, CA: Rand Research Association, 1977.

1127. Lopez, D.E. "Chicano Language Loyalty in an Urban Setting." *Sociological and Social Research; an International Journal* 62 (January 1978): 267–278.

1128. Maldonado, E. "Contract Labor and the Origins of Puerto Rican Communities in the United States." *International Migration Review* 13, 1 (1979): 103–121.

1129. Massey, D.S. "Residential Segregation of Spanish-Americans in United States Urbanized Areas." *Demography* 16, 4 (November 1979): 553–563.

1130. Metzgar, J.V. "Guns and Butter: Albuquerque Hispanics, 1940–1975." *New Mexico Historical Review* 56, 2 (April 1981): 117–140.

1131. Miranda, G.E. "Gente de Razon Marriage Patterns in Spanish and Mexican California: A Case Study of Santa Barbara and Los Angeles." *Southern California Quarterly* 63 (Spring 1981): 1–21.

1132. Mirande, A. "Fear of Crime and Fear of the Police in a Chicano Community." *Sociology and Social Research; an International Journal* 64 (July 1980): 528–541.

1133. Morley, F., and C. Hawk. "Abilene and the Spanish-Speaking Community." *Public Management* 63, 1 (January/February 1981): 27–28.

1134. Pastor, J. "Hispanic Female in City Management." *Public Management* 62, 1 (October 1980): 20–21.

1135. Romo, R. *East Los Angeles: History of a Barrio.* Austin, TX: University of Texas Press, 1983.

1136. Ropka, G.M. *The Evolving Residential Patterns of the Mexican, Puerto Rican, and Cuban Population in the City of Chicago.* New York: Arno Press, 1980. Reprint.

1137. Rosales, F.A. "Regional Origins of Mexican Immigrants to Chicago during the 1920's." *Aztlan* 7, 1 (Summer 1976): 187–201.

1138. Ross, Elmer L. *Factors in Residence Patterns among Latin-Americans in New Orleans, Louisiana.* New York: Arno Press, 1980. Reprint.

1139. Salcido, R.M. "Problems of the Mexican-American Elderly in an Urban Setting." *Social Casework* 60, 1 (December 1979): 609-615.

1140. Sepulveda, C. "Una Colonia de Obreros: East Chicago, Indiana." *Aztlan* 7, 1 (Summer 1976): 307-326.

1141. Shannon, L. *Minority Migrants in the Urban Community: Mexican-American and Negro Adjustments to Industrial Society.* Beverly Hills, CA: Sage Publications, 1972.

1142. Shannon, L.W., and V.F. Davison. "Change in the Economic Absorption of Immigrant Mexican-American and Negroes in Racine, Wisconsin between 1960-1971." *International Migration Review* 11 (Summer 1977): 190-214.

1143. Taylor, P.S. "Mexican Women in Los Angeles Industry in 1928." *Aztlan* 11 (Spring 1980): 99-131.

1144. Tucey, M., and D. Hornbeck. "Anglo Immigration and the Hispanic Town: A Study of Urban Change in Monterey, California 1835-1900." *Social Science Journal* 13, 2 (April 1976): 1-8.

1145. Williamson, D. "Adaptation to Sociocultural Change: Working-Class Cubans in New Orleans." *Caribbean Studies* 16, 3/4 (October/January 1977): 217-227.

1146. Wilson, K.L., and A. Portes. "Immigrant Enclaves: An Analysis of the Labor Market Experiences of Cubans in Miami." *American Journal of Sociology* 86, 2 (September 1980): 295-319.

1147. Winsberg, M.D. "Housing Segregation of a Predominantly Middle-Class Population: Residential Patterns of Development by the Cuban Immigration into Miami 1950-1970." *American Journal of Economics and Sociology* 38, 4 (October 1979): 403-418.

Hungarian

1148. Schuchat, M.G. "Hungarian-Americans in the Nation's Capital." *Anthropological Quarterly* 54, 2 (April 1981): 89-93.

1149. Weinberg, D.E. "Ethnic Identity in Industrial Cleveland: The Hungarians 1900-1920." *Ohio History* 86, 3 (Summer 1977): 171-186.

Irish

1150. Burchell, R.A. "Irish of San Francisco, 1848-1880." *Current Anthropology* 19, 3 (June 1978): 458.

1151. ————. *The San Francisco Irish 1848-1880*. Berkeley, CA: University of California Press, 1980.

1152. Clark, D. "Babies in Bondage: Indentured Irish Children in Philadelphia in the Nineteenth Century." *Pennsylvania Magazine of History and Biography* 10, 4 (October 1977): 475-486.

1153. ————. *The Irish in Philadelphia: Ten Generations of Urban Experience*. Philadelphia, PA: Temple University Press, 1973.

1154. Erie, S.P. "Politics, the Public Sector and Irish Social Mobility: San Francisco, 1870-1900." *Western Political Quarterly* 31, 2 (June 1978): 274-289.

1155. Fanning, C. *Mr. Dooley and the Chicago Irish: An Anthology*. New York: Arno Press, 1976. Reprint.

1156. Funchion, M.F. *Chicago's Irish Nationalists, 1881-1890*. New York: Arno Press, 1976. Reprint.

1157. ————. "Irish Nationalists and Chicago Politics in the 1880's." *Erie-Ireland* 10 (Summer 1975): 3-19.

1158. Gabriel, R.A. *The Irish and Italians: Ethnics in City and Suburb*. New York: Arno Press, 1980. Reprint.

1159. Groreman, C. "Working-Class Immigrant Women in Mid-Nineteenth-Century New York: The Irish Women's Experience." *Journal of Urban History* 4, 3 (May 1978): 255-274.

1160. Monahan, K. "The Irish Hour: An Expression of the Musical Taste and the Cultural Values of the Pittsburgh Irish Community." *Ethnicity* 4, 3 (September 1977): 207-215.

1161. Silver, C. "A New Look at Old South Urbanization: The Irish Workers in Charleston, South Carolina 1840-1860." *South Atlantic Urban Studies* 3 (1979): 141-172.

1162. Vinyard, J.E. *The Irish on the Urban Frontier: Detroit 1850-1880*. New York: Arno Press, 1976. Reprint.

1163. Walsh, V.A. "Fanatic Heart: The Cause of Irish-American Nationalism in Pittsburgh during the Gilded Age." *Journal of Social History* 15, 2 (Winter 1981): 187-204.

1164. Wilentz, R.S. "Industrializing America and the Irish:
       Towards the New Departure." *Labor History* 20, 4
       (1979): 579-595.

                                Italian

1165. Baccari, Alessandro, and A.M. Canepa. "The Italians
       of San Francisco in 1865: G.B. Cerrutis' Report to
       the Ministry of Foreign Affairs." *California His-
       tory* 60 (Winter 1981-1982): 350-369.

1166. Briggs, J.W. *An Italian Passage: Immigrants to Three
       American Cities 1890-1930.* New Haven, CT: Yale Uni-
       versity Press, 1978.

1167. Cannistraro, P.V. "Facism and Italian-Americans in
       Detroit 1933-1935." *International Migration Review*
       9, 1 (1975): 29-40.

1168. Della Femina, J. *An Italian Grows in Brooklyn.* Boston,
       MA: Little, Brown, 1978.

1169. Ferroni, D.C. *The Italians in Cleveland: A Study in
       Assimilation.* New York: Arno Press, 1980. Reprint.

1170. Gumina, D.P. *The Italians in San Francisco, 1850-1930.*
       Staten Island, NY: Center for Migration Studies, 1977.

1171. Harney, R.F., and J.V. Scarpaci. *Little Italies in
       North America.* Toronto: Multicultural History
       Society of Ontario, 1981.

1172. Hesse-Biber, S. "The Ethnic Ghetto and Private Wel-
       fare: A Case Study of Southern Italian Immigration
       to the United States, 1880-1914." *Urban and Social
       Change Review* 12, 2 (1979): 9-15.

1173. Mangione, J. *Mount Allegro: A Memoir of Italian-
       American Life.* New York: Columbia University Press,
       1981.

1174. Mondello, S.A. *The Italian Immigrant in Urban America,
       1880-1920, As Reported in the Contemporary Periodical
       Press.* New York: Arno Press, 1980. Reprint.

1175. Mormino, G.R. "A House on the Hill: Mobility Patterns
       in an Italian Neighborhood." *The Maryland Historian*
       11, 2 (Fall 1980): 13-24.

1176. Mormino, G. "Immigrant Editor: Making a Living in
       Urban America." *Journal of Ethnic Studies* 9, 1
       (Spring 1981): 81-85.

1177. Peroni, Peter. *The Burg: An Italian-American Community at Bay in Trenton.* Washington, D.C.: University Press of America, 1979.

1178. Pozetta, G.E. "The Italian Immigrant Press of New York City: The Early Years, 1880-1915." *Journal of Ethnic Studies* 1, 3 (Fall 1973): 32-46.

1179. ————. "Italians and Urban America." *Journal of Urban History* 6, 3 (May 1980): 357-365.

1180. Scarpaci, J.A. *Italian Immigrants in Louisiana's Sugar Parishes: Recruitment, Labor Conditions and Community Relations 1880-1910.* New York: Arno Press, 1980. Reprint.

1181. Scheimi, R.D. *The Italian Community of San Francisco: A Descriptive Study.* New York: Arno Press, 1980.

1182. Thompson, B. *Cultural Ties as Determinants of Immigrant Settlement in Urban Areas: A Case Study of the Growth of an Italian Neighborhood in Worcester, Massachusetts, 1875-1922.* New York: Arno Press, 1980. Reprint.

1183. Yans-McLaughlin, V. *Family and Community: Italian Immigrants in Buffalo 1880-1977.* Ithaca, NY: Cornell University Press, 1977.

Jewish

1184. Bernstein, S. "The Economic Life of the Jews in San Francisco during the 1860's as Reflected in the City Directories." *American Jewish Archives* 27, 1 (April 1975): 70-78.

1185. Bloom, L.A. "A Successful Jewish Boycott of the New York City Public Schools--Christmas 1906." *American Jewish History* 70, 2 (December 1980): 180-188.

1186. Cohn, J.M. "Demographic Studies of Jewish Communities in the United States: A Bibliographic Introduction and Survey." *American Jewish Archives* 32, 1 (April 1980): 35-51.

1187. Davis, D. "A Geographic Perspective of Jewish Attitudes towards Neighborhood Transition: A Core Study." *Geographic Perspectives* 45 (Spring 1980): 35-48.

1188. Decker, P.R. "Jewish Merchants in San Francisco: Social Mobility on the Urban Frontier." *American Jewish History* 68, 4 (June 1979): 396-407.

1189. Dubrovsky, G. "Growing Up in Farmingdale." *American Jewish History* 71 (December 1981): 239-256.

1190. Dwork, D. "Health Conditions of Immigrant Jews on the Lower East Side of New York 1880-1914." *Medical History* 25, 1 (June 1981): 1-40.

1191. Feldman, J.S. "The Pioneers of a Community: Regional Diversity Among the Jews of Pittsburgh, 1845-1861." *American Jewish Archives* 32, 2 (November 1980): 119-124.

1192. Fierman, F.S. "The Schwartz Family of El Paso: The Story of a Pioneer Jewish Family in the Southwest." *Southwest Studies* 61 (1980): 5-76.

1193. Fredman, R.G. "Cosmopolitans at Home: The Sephardic Jews of Washington, D.C." *Anthropological Quarterly* 54, 2 (April 1981): 61-67.

1194. Fried, L. "Jacob Riis and the Jews: The Ambivalent Quest for Community." *American Studies* 20, 1 (1979): 5-14.

1195. Gelfand, M.B. "Progress and Prosperity: Jewish Social Mobility in Los Angeles in the Booming Eighties." *American Jewish History* 68, 4 (June 1979): 408-433.

1196. Gurock, J. *When Harlem Was Jewish, 1870-1930.* New York: Columbia University Press, 1979.

1197. Heimouks, R.B. *The Chicago Jewish Source Book.* Chicago, IL: Follet Publishing Co., 1981.

1198. Hertzberg, S. "The Jewish Community of Atlanta from the End of the Civil War until the Eve of the Frank Case." *American Jewish Historical Quarterly* 62, 3 (March 1973): 250-285.

1199. Howe, I. *The Immigrant Jews of New York 1881 to the Present.* Boston, MA: Routledge and Kegan Paul, 1976.

1200. Jaret, C. "Recent Patterns of Chicago Jewish Residential Mobility." *Ethnicity* 6, 3 (September 1979): 235-248.

1201. Kartman, L.L. "Jewish Occupational Roots in Baltimore at the Turn of the Century." *Maryland Historical Magazine* 74, 1 (Spring 1979): 52-61.

1202. Livingston, J. "The Industrial Removal Office: The Galveston Project and the Denver Jewish Community." *American Jewish History* 68, 4 (June 1979): 434-458.

1203. Mayer, E. *From Suburb to Shtetl: The Jews of Boro Park.* Philadelphia, PA: Temple University Press, 1979.

1204. Moore, D.D. *At Home in America: Second Generation New York Jews.* New York: Columbia University Press, 1981.

1205. Pierce, L.E. "The Jewish Settlement on St. Paul's Lower West Side." *American Jewish Archives* 28 (November 1976): 143-161.

1206. Raphael, M.L. "Federated Philanthropy in an American Jewish Community 1904-1908." *American Jewish History* 68, 2 (December 1978): 147-162.

1207. ————. *Jews and Judaism in a Mid-Western Community: Columbus, Ohio.* Columbus, OH: Columbus Historical Society, 1979.

1208. Rischin, M. *The Promised City: New York's Jews 1870-1914.* Cambridge, MA: Harvard University Press, 1977.

1209. Rockaway, R.A. "Antisemitism in an American City: Detroit 1850-1914." *American Jewish Historical Quarterly* 64, 1 (September 1974): 42-54.

1210. Rockaway, R.A. "The Future of American Jewish Communal Histories." *Journal of Urban History* 7 (February 1981): 255-260.

1211. Romanofsky, P. "To Save ... Their Souls ...: The Care of Dependent Jewish Children in New York City 1900-1905." *Journal of Jewish Studies* 36, 3 (July/October 1974): 253-261.

1212. Shankman, A. "Atlanta Jewry 1900-1930." *American Jewish Archives* 25, 1 (April 1973): 131-155.

1213. Silverman, M. "Class Kinship and Ethnicity: Patterns of Jewish Upward Mobility in Pittsburgh, Pennsylvania." *Urban Anthropology* 7, 1 (Spring 1978): 25-44.

1214. Toll, W. "Fraternalism and Community Structure on the Urban Frontier: The Jews of Portland, Oregon--A Case Study." *Portland Historical Review* 47, 3 (August 1978): 369-404.

1215. ————. "Mobility, Fraternalism and Jewish Cultural Change: Portland, Oregon, 1910-1930." *American Jewish History* 68, 4 (June 1979): 459-491.

1216. ————. "Volunteerism and Modernization in Portland Jewry: The B'nai B'rith in the 1920's." *Western Historical Quarterly* 10, 1 (January 1979): 21-38.

1217. Varady, David P. "Migration and Mobility Patterns of the Jewish Population of Cincinnati." *American Jewish Archives* 32, 1 (April 1980): 78-88.

Native American

1218. Bohland, J.R. "Indian Residential Segregation in the Urban Southwest: 1970-1980." *Social Science Quarterly* 63, 4 (1982): 749-761.

1219. Brenner, E.M. "To Pray or To Be Prey: That Is The Question Strategies for Cultural Autonomy of Massachusetts Praying Town Indians." *Ethnohistory* 27, 2 (Spring 1980): 135-152.

1220. Fiske, S.J. "Urban Indian Institutions: A Reappraisal from Los Angeles." *Urban Anthropology* 8, 2 (Summer 1979): 149-172.

1221. Hansen, K.T. "Ethnic Group Policy and the Politics of Sex: The Seattle Indian Case." *Urban Anthropology* 8, 1 (1979): 29-48.

1222. Morgan, A. "Indians and Immigrants: A Comparison of Groups New to the City." *Journal of Ethnic Studies* 4, 4 (Winter 1977): 17-28.

1223. Phillips, G.H. "Indians in Los Angeles 1781-1975: Economic Integration, Social Disintegration." *Pacific Historical Review* 49, 3 (1980): 427-451.

1224. Red Horse, J.G. "Family Behavior of Urban American Indians." *Social Casework* 59, 2 (February 1978): 67-72.

1225. Sorkin, A.L. *The Urban American Indian*. Lexington, MA: Lexington Books, 1979.

1226. Strauss, J.H., et al. "An Experimental Outreach Legal Aid Program for an Urban Native American Population Utilizing Legal Paraprofessionals." *Human Organization* 38, 4 (1979): 386-394.

1227. Strauss, J.H., and B.A. Chadwick. "Urban Indian Adjustment." *American Indian Culture and Research Journal* 3, 2 (1979): 23-38.

1228. Tax, S. "Impact of Urbanization on American Indians." *American Academy of Political and Social Science Annals* 436 (March 1978): 121-136.

1229. Walker, R.D. "Treatment Strategies in an Urban Indian Alcoholism Program." *Journal of Studies on Alcohol* 9, 1 (January 1981): 171-184.

Portuguese

1230. Leder, H.H. *Cultural Persistence in a Portuguese American Community*. New York: Arno Press, 1980. Reprint.

Scandinavian

1231. Anderson, H.H. "Scandinavian Immigration in Milwaukee Naturalization Records." *Milwaukee History* 1 (Spring/ Summer 1978): 25-37.

1232. *The Finnish Experience in the Western Great Lakes Region: New Perspectives*. St. Paul, MN: Minnesota University Immigration History Research, 1976.

1233. Webster, J.R. "Domestication and Americanization: Scandinavian Women in Seattle 1888 to 1900." *Journal of Urban History* 4, 3 (May 1978): 275-290.

Slavic

1234. Baskauska, L. "Multiple Identities: Adjusted Lithuanian Refugees in Los Angeles." *Urban Anthropology* 6, 2 (Summer 1977): 141-154.

1235. Bennett, L.A. "Washington and Its Serbian Emigrés: A Distinctive Blend." *Anthropological Quarterly* 54, 2 (April 1981): 82-88.

1236. Bigelow, B. "Marital Assimilation of Polish-Catholic Americans: A Case Study in Syracuse, N.Y.; 1940-1970." *Professional Geographer* 32, 4 (November 1980): 431-438.

1237. Boatler, J. "Patterns of Infant Mortality in the Polish Community of Chappell Hill, Texas, 1895-1944." *Human Biology* 55, 1 (1983): 9-18.

1238. Bukowczyk, J.J. "The Immigrant 'Community' Reexamined: Political and Economic Tensions in a Brooklyn Polish Settlement 1884-1894." *Polish American Studies* 37, 2 (Autumn 1980): 5-16.

1239. Crozier, W.L. "A People Apart: A Census Analysis of the Polish Community of Winona Minnesota 1880-1905." *Polish American Studies* 38 (Spring 1981): 5-22.

1240. Cuba, S.L. "A Polish Community in the Urban West: St. Joseph's Parish in Denver, Colorado." *Polish-American Studies* 36 (Spring 1979): 33-74.

1241. Greenbaum, P.E., and S.D. Greenbaum. "Territorial Per-
       sonalization: Group Identity and Social Interaction
       in a Slavic-American Neighborhood." *Environment and
       Behavior* 13, 5 (September 1981): 57-89.

1242. Kuzniewski, A.J. "Milwaukee's Poles 1866-1918: The
       Rise and Fall of a Model Community." *Milwaukee His-
       tory* 1, 1/2 (1978): 13-24.

1243. Lichtenstein, G. "A Little Russia Grows in Brooklyn."
       New York Magazine 44 (June 1981): 30-34.

1244. Obidinski, E.E. *Ethnic to Status Group: A Study of
       Polish Americans in Buffalo*. New York: Arno Press,
       1981. Reprint.

1245. Parot, J. "The Racial Dilemma in Chicago's Polish
       Neighborhoods, 1920-1979." *Polish American Studies*
       32 (Autumn 1975): 27-38.

1246. Pienkos, D. "Politics, Religion and Change in Polish
       Milwaukee 1900-1930." *Wisconsin Magazine of History*
       61, 3 (Spring 1978): 179-209.

1247. Sadler, C. "Political Dynamite: The Chicago Polonia
       and President Roosevelt in 1944." *Journal of Illinois
       State Historical Society* 71, 2 (1978): 119-132.

1248. Stevenson, J.C., and R.J. Duquesnoy. "Distribution of
       HLA Antigens in Polish and German Populations in Mil-
       waukee, Wisconsin." *Physical Anthropology* 60 (Jan-
       uary 1979): 19-23.

1249. Wrobel, P. *Our Way: Family, Parish and Neighborhood in
       a Polish-American Community*. Notre Dame, IN: Notre
       Dame University Press, 1979.

Welsh

1250. Davies, P.G. "Touring the Welsh Settlements of South
       Dakota, 1891." *South Dakota History* 10 (Summer 1980):
       223-240.

1251. Ellis, D.M. "The Assimilation of the Welsh in Central
       New York." *Western Historical Review* 4, 4 (October
       1973): 424-447.

"SEWER" SOCIALISM

As a political force, socialism has never gained more than
a small number of adherents in the United States. This record
of failure is well documented in the literature. A recent
analysis is found in Harvey Klehr's "Leninism, Lewis Corey,
and the Failure of American Socialism" (*Labor History* 18, 2
(1977): 249-256). However, within any description of this
unaccomplished political success, socialism can be found to
be successful in the municipal setting, specifically Milwaukee
(Wisconsin), Bridgeport (Connecticut), and Reading (Pennsylvania) which were centers for socialist political success.
Municipal socialism enjoyed a golden period during the first
third of the twentieth century within the American political
spectrum.

In 1910, Milwaukee elected socialist Emil Seidel as mayor.
In addition, the voters of Milwaukee provided Seidel with a
socialist-dominated administration: city attorney, city
treasurer, city comptroller, twenty-one of thirty-five alder-
men, and one-fourth of the school board were socialists. This
was the largest electoral victory socialism has ever enjoyed
in the United States. Frederick Olson's "Milwaukee's First
Socialist Administration 1910-1912: A Political Evaluation"
(*Mid-America* 43, 2 (1961): 197-207), presents an analysis of
the socialist success and its meaning for that particular
time period in the history of Milwaukee and the state of Wisconsin. From 1927 to 1935, the Socialist Party was the dominant
political force in Reading, Pennsylvania. The mayor of Reading
was a socialist in 1927 and 1935. Henry Stetler's *The Socialist Movement in Reading, Pennsylvania 1896-1936* (1974) is useful narrative for charting the success and accomplishments of
socialist politicians in Reading. In Bridgeport, Connecticut,
socialism prospered from 1933 to 1957 under socialist mayor
Jasper McLevy and his followers.

Municipal socialism found favorable support through its
efforts to reform city finances, establish public ownership of
utilities and transportation, and curb the hostility of the
police toward workers and their efforts to unionize. But this
"sewer" socialism, so-called due to socialist efforts to curb
public utilities, never acquired a permanent base of support.

Socialism in Reading and Milwaukee faded during the late
1930s. The Bridgeport socialists held power into the 1950s.
These few isolated cases constitute the record of socialist
success. The failure of socialism as a political force is
also due, in part, to the success attained in Milwaukee,
Reading, and Bridgeport. Successful practical solutions to
municipal problems begun by socialists once elected to municipal office helped maintain elected socialists in power
within municipal governments but also convinced other political
parties of the need to adopt similar practical solutions.
There was never any attempt by socialists to translate their
successful practical projects in municipal settings into long-
range and ideological success. New Deal Democrats quickly
adopted many socialist and other reform ideas in efforts to
lessen the effects of the Great Depression. Socialist goals--
nationalization of the means of production and distribution,
democratic planning, and production for use rather than profit--
were never accepted by the American electorate. Socialists
who did manage to get elected to municipal offices never ac-
complished more than their practical reform goals for fear of
losing elections. The need to retain political power limited
and defined the scope of socialism within the municipal setting.

Research into the influence of socialism in American local
governing bodies continues. Two interesting works have been
published examining socialist influence in New York City.
Charles Leinenweber's "The Class and Ethnic Bases of New York
City Socialism, 1904-1915" (*Labor History* 22, 1 (1981): 31-56)
and H. Perrier's "Socialists and the Working Class in New
York: 1890-1896" (*Labor History* 22, 4 (1981): 485-511) both
analyze the class basis for the socialist movement within the
city. Richard Oestreicher's "Socialism and the Knights of
Labor in Detroit, 1877-1886" (*Labor History* 22, 1 (1981): 5-30)
provides a similar study for Detroit.

The following books and journal articles provide a representative list of research conducted into the influence of
socialism in American cities. This list is a starting point
for further investigation and discussion.

General

1252. Bassett, M. "Municipal Reform and the Socialist Party,
        1910-1914." *Australian Journal of Political History*
        19, 2 (August 1973): 179-187.

1253. Bolino, A.C. "American Socialism; Flood and Ebb: The
        Use and Decline of the Socialist Party in America,
        1901-1912." *American Journal of Economics and
        Sociology* 22, 2 (April 1963): 287-301.

1254. Cantor, M. *Divided Left; American Radicalism, 1900-1975.* New York: Farrar, Straus and Giroux, 1978.

1255. Douglas, P.H. "The Socialist Vote in the Municipal Elections of 1917." *National Municipal Review* 6, 3 (March 1918): 131-138.

1256. Duram, J.C. "Algernon Lee's Correspondence with Karl Kautsky: An "Old Guard." Perspectives on the Failure of American Socialism." *Labor History* 20, 3 (Summer 1979): 420-434.

1257. Egbert, D.D., and S. Persons, eds. *Socialism and American Life.* 2 vols. Princeton, NJ: Princeton University Press, 1952.

1258. French, R.A., and F.E.I. Hamilton. *The Socialist City: Spatial Structure and Urban Policy.* New York: John Wiley and Sons, 1980.

1259. Hays, Samuel P. "The Politics of Reform in Municipal Government in the Progressive Era." *Pacific Northwest Quarterly* 55, 4 (October 1964): 157-169.

1260. Hereshaff, D. *American Disciples of Marx: From the Age of Jackson to the Progressive Era.* Detroit, MI: Wayne State University Press, 1967.

1261. Hillquit, M. *History of Socialism in the United States.* New York: Funk and Wagnalls, 1910.

1262. Horn, M. *Intercollegiate Socialist Society 1905-1921; Origins of the Modern Student Movement.* Boulder, CO: Westview Press, 1979.

1263. Kipnis, I. *The American Socialist Movement, 1897-1912.* New York: Columbia University Press, 1952.

1264. Klehr, H. "Leninism, Lewis Corey, and the Failure of American Socialism." *Labor History* 18, 2 (Spring 1977): 249-256.

1265. Leinenweber, C. "The American Socialist Party and New Immigrants." *Science and Society* 32, 4 (Winter 1968): 1-25.

1266. Lipow, A. *Authoritarian Socialism in America; Edward Bellamy and the Nationalist Movement.* Berkeley, CA: University of California Press, 1982.

1267. Macy, J. *Socialism in America.* Garden City, NY: Doubleday, 1916.

1268. Martin, R.L. *Fabian Freeway; High Road to Socialism in the USA 1884-1966.* Boston, MA: Western Islands, 1966.

1269. Moore, R.L. *Emergence of an American Left; Civil War to World War I.* New York: John Wiley and Sons, 1973.

1270. Morgan, H.W. *American Socialism, 1900-1960.* Englewood Cliffs, NJ: Prentice-Hall, Inc., 1964.

1271. Newton, B. "Henry George and Henry M. Hyndman, II: The Erosion of the Radical-Socialist Coalition, 1884-1889." *American Journal of Economic Sociology* 36, 3 (July 1977): 311-321.

1272. Nord, D.P. "The Appeal to Reason and American Socialism, 1901-1920." *Kansas History* 1, 2 (Summer 1978): 75-89.

1273. Proefriedt, W. "Socialist Criticisms of Education in the United States: Problems and Possibilities." *Harvard Education Review* 50, 4 (November 1980): 467-480.

1274. Quint, H.H. *The Forging of American Socialism; Origins of the Modern Movement.* Columbia, SC: University of South Carolina Press, 1953.

1275. Reynolds, R.D., Jr. "Pro-War Socialists: Intolerant or Blood Thirsty?" *Labor History* 17, 3 (Summer 1976): 413-415.

1276. Seretan, L.G. "Daniel DeLeon as American." *Wisconsin Magazine of History* 61, 3 (Spring 1978): 210-223.

1277. Shannon, D.A. "The Socialist Party before the First World War: An Analysis." *Mississippi Valley Historical Review* 38, 2 (September 1951): 279-288.

1278. Stave, B.M. *Socialism and the Cities.* Port Washington, NY: Kennikat Press, 1975.

1279. Stevenson, J. "Daniel DeLeon and European Socialism, 1890-1914." *Science and Society* 44, 2 (Summer 1980): 199-223.

1280. Stow, R.N., ed. "Conflict in the American Socialist Movement, 1897-1901: A Letter from Thomas J. Morgan to Henry Demarest Lloyd, July 18, 1901." *Journal of the Illinois State Historical Society* 71, 2 (May 1978): 133-142.

1281. Waite, J.A. *Masses: 1911-1917; a Study in American Rebellion.* College Park, MD: University of Maryland Press, 1951.

1282. Warren, F.A. *An Alternative Vision: The Socialist Party in the 1930's.* Bloomington, IN: Indiana University Press, 1974.

1283. Weinstein, J. *The Decline of Socialism in America, 1912-1925*. New York: Monthly Review Press, 1967.

1284. Yellowitz, I. "Morris Hillquit: American Socialism and Jewish Concerns." *American Jewish History* 68, 2 (December 1978): 163-188.

The East

Massachusetts

1285. Bedford, H.F. "The Haverhill Social-Democrat: Spokesman for Socialism." *Labor History* 2, 1 (Winter 1961): 82-89.

1286. ————. *Socialism and the Workers in Massachusetts 1886-1912*. Amherst, MA: University of Massachusetts Press, 1966.

1287. ————. "The Socialist Movement in Haverhill." *Essex Institute Historical Collection* 99, 1 (January 1963): 33-47.

1288. Faler, P.G. *Workingmen, Mechanics, and Social Change in Lynn, Massachusetts, 1800-1860*. Unpublished dissertation. University of Wisconsin, 1971.

New Jersey

1289. Ebner, M.H. *The Historian's Passaic, 1855-1912: City Building in Post Civil War America*. Unpublished dissertation. University of Virginia, 1974.

1290. ————. "The Passaic Strike of 1912 and the Two IWW's." *Labor History* 11, 4 (Fall 1970): 452-466.

1291. Levin, H. "The Paterson Silkworkers' Strike of 1913." *King's Crown Essays* 9 (Winter 1961): 44-64.

New York

1292. Dubofsky, M. "Success and Failure of Socialism in New York City, 1900-1918: A Case Study." *Labor History* 9, 4 (Fall 1968): 361-375.

1293. Gaffield, C. "Big Business, the Working Class, and Socialism in Schenectady, 1911-1916." *Labor History* 19, 3 (Summer 1978): 350-372.

1294. Gorenstein, A. "A Portrait of Ethnic Politics: The Socialists and the 1908 and 1910 Congressional Elections of the West Side (Manhattan Island)." *American Jewish Historical Quarterly* 50, 3 (March 1961): 202-238.

1295.   Isserman, M.   "Inheritance Lost: Socialism in Rochester
        1917-1919."   *Rochester History* 39, 4 (October 1977):
        1-24.

1296.   Josephson, H.   "Dynamics of Repression: New York during
        the Red Scare."   *Mid-America* 59, 3 (October 1977):
        131-146.

1297.   Leinenweber, C.   "The Class and Ethnic Bases of New
        York City Socialism, 1904-1915."   *Labor History* 22, 1
        (Winter 1981): 31-56.

1298.   ———.   "Socialists in the Streets: The New York City
        Socialist Party in Working-Class Neighborhoods, 1908-
        1918."   *Science and Society* 41, 2 (Summer 1977): 152-
        171.

1299.   Perrier, H.   "Socialists and the Working Class in New
        York: 1890-1896."   *Labor History* 22, 4 (Fall 1981):
        485-511.

1300.   Powell, E.H.   "Reform, Revolution, and Reaction: A Case
        of Organized Conflict," in I. Horowitz, ed., *The New
        Sociology*.   New York: Oxford University Press, 1964.

Pennsylvania

1301.   Bonosky, P.   "The Story of Ben Careathers."   *Masses
        and Mainstream* 6, 7 (July 1953): 34-44.

1302.   Foner, P.S.   "Caroline Hollingsworth Pemberton: Phila-
        delphia Socialist Champion of Black Equality."   *Penn-
        sylvania History* 43, 3 (July 1976): 227-251.

1303.   Ford, R.   *Germans and Other Foreign Stock: Their Part
        in the Evolution of Reading, PA*.   Unpublished dis-
        sertation.   University of Pennsylvania, 1963.

1304.   Hendrickson, K.E.   "The Socialists of Reading, Penn-
        sylvania and World War I--A Question of Loyalty."
        Pennsylvania History 36, 4 (October 1969): 430-450.

1305.   ———.   "Triumph and Disaster: The Reading Socialists
        in Power and Decline, 1932-1939, Part II."   *Pennsyl-
        vania History* 40, 4 (October 1973): 381-411.

1306.   Hodges, H.G.   "Socialists Lose Control in Reading,
        Pennsylvania."   *National Municipal Review* 20, 12
        (December 1931): 743-744.

1307.   ———.   "Four Years of Socialism in Reading, Penn-
        sylvania."   *National Municipal Review* 20, 5 (May
        1931): 281-289.

1308. Maurer, J.H. *It Can Be Done.* New York: Rand School,
       1938.

1309. Pratt, W.C. *The Reading Socialist Experience: A Study
       in Working-Class Politics.* Unpublished dissertation.
       Emory University, 1969.

1310. ————. "Women and American Socialism: The Reading
       Experience." *Pennsylvania Magazine of History and
       Biography* 99, 1 (January 1975): 72-91.

1311. Stetler, H.G. *The Socialist Movement in Reading, Penn-
       sylvania 1896-1936.* Philadelphia, PA: Porcupine
       Press, 1974.

                            The Midwest

1312. Puotinen, A.E. *Finnish Radicals and Religion in Mid-
       western Mining Towns 1865-1914.* New York: Arno
       Press, 1979.

Illinois

1313. Buenker, J.D. "Illinois Socialists and Progressive
       Reform." *Journal of the Illinois State Historical
       Society* 63, 4 (Winter 1970): 368-386.

1314. Stevens, E.W. "The Socialist Party of America in Muni-
       cipal Politics: Canton, Illinois, 1911-1920." *Journal
       of the Illinois State Historical Society* 72, 2 (Novem-
       ber 1979): 257-272.

Michigan

1315. Holli, M.G. *Reform in Detroit: Hazen S. Pingree and
       Urban Politics.* New York: Oxford University Press,
       1969.

1316. Oestreicher, R. "Socialism and the Knights of Labor
       in Detroit, 1877-1886." *Labor History* 22, 1 (Winter
       1981): 5-30.

Minnesota

1317. Cannon, J.P. *Socialism on Trial.* New York: Pioneer,
       1944.

1318. Nord, D.P. "Minneapolis and the Pragmatic Socialism
       of Thomas Van Lear." *Minnesota History* 45, 1 (Spring
       1976): 2-10.

1319. ————. *Socialism in One City: A Political Study of*

Minneapolis in the Progressive Era. Unpublished
dissertation. University of Minnesota, 1972.

Wisconsin

1320. Attoe, W., and M. Latus. "The First Public Housing:
       Sewer Socialism's Garden City for Milwaukee."
       Journal of Popular Culture 10, 1 (Summer 1976):
       142-149.

1321. Collier, J. "Experiment in Milwaukee." *Harper's
       Weekly* 55 (August 12, 1911): 11.

1322. England, G.A. "Milwaukee's Socialist Government."
       American Review of Reviews 42 (November 1910): 445-
       455.

1323. Ettenheim, S.C. *How Milwaukee Voted 1848-1968.* Mil-
       waukee, WI: University of Wisconsin Press, 1970.

1324. Hoan, D.W. *City Government: The Record of the Milwaukee
       Experiment.* New York: Harcourt, Brace and Company,
       1936.

1325. Howe, F.C. "Milwaukee: A Socialist City." *Outlook*
       90 (June 25, 1910): 411-421.

1326. Korman, G. *Industrialization, Immigration, and Ameri-
       canization: The View from Milwaukee, 1886-1921.*
       Madison, WI: University of Wisconsin Press, 1967.

1327. Lorence, J.J. "The Milwaukee Connection: The Urban-
       Rural Link in Wisconsin Socialism, 1910-1920."
       Milwaukee History 3 (Winter 1980): 102-111.

1328. Miller, S.M. *Victor Berger and the Promise of Construc-
       tive Socialism.* Westport, CT: Greenwood Press, 1973.

1329. Olson, F.I. *The Milwaukee Socialists, 1897-1941.*
       Dissertation: Harvard Univ. Press, 1952.

1330. ————. "Milwaukee's First Socialist Administration,
       1910-1912: A Political Evaluation." *Mid-America*
       43, 2 (July 1961): 197-207.

1331. ————. "The Socialist Party and the Union in Mil-
       waukee, 1900-1912." *Wisconsin Magazine of History*
       44, 2 (Winter 1961): 110-116.

1332. Reese, W.J. "Partisans of the Proletariat: The Social-
       ist Working Class and the Milwaukee Schools, 1890-
       1920." *History of Education Quarterly* 21, 1 (Spring
       1981): 3-50.

1333. Reinders, R.C. "Daniel W. Hoan and the Milwaukee
       Socialist Party during the First World War." *Wisconsin Magazine of History* 36, 3 (Autumn 1952): 48-55.

1334. "Socialists Elect Milwaukee's Mayor." *Survey* 36
       (April 15, 1916): 69-70.

1335. Stern, E.C. "The Non-Partisan Election Law: Reform
       or Anti-Socialism." *Historical Messenger* 16, 1
       (1960): 8-11.

1336. Thelen, D.P. *The New Citizenship: Origins of Progressivism in Wisconsin, 1885-1900.* Columbia, MO:
       University of Missouri Press, 1972.

1337. Velicer, L. *Municipal Ownership and the Manitowoc,
       Wisconsin Socialists, 1905-1917.* Manitowoc, WI:
       Manitowoc County Historical Society, 1978.

1338. "Victor Berger, the Organizer of the Socialist Victory
       in Milwaukee." *Current Literature* 49 (September
       1910): 265-269.

1339. Wachman, M. *History of the Social-Democratic Party
       of Milwaukee, 1897-1910.* Urbana, IL: University of
       Illinois Press, 1945.

1340. Zeiner, F.P. "Dan Norca, Successful Mayor (Milwaukee)."
       Historical Messenger 17, 1 (March 1961): 21-25.

The South

Louisiana

1341. Tregle, J.G., Jr. "Thomas J. Durant, Utopian Socialism
       and the Failure of Presidential Reconstruction in
       Louisiana." *Journal of Social History* 45, 1 (November
       1979): 485-512.

West Virginia

1342. Barkley, F.A. *The Socialist Party in West Virginia
       from 1898-1920: A Study in Working Class Radicalism.*
       Unpublished dissertation. University of Pittsburgh,
       1971.

The West

1343. Green, J.R. *Socialism and the Southwestern Class
       Struggle, 1898-1918: A Study of Radical Movements*

in Oklahoma, Texas, Louisiana and Arkansas. Unpublished dissertation. Yale University, 1972.

California

1344. Hines, T.S. "Housing, Baseball, and Creeping Socialism: The Battle of Chavez Ravine, Los Angeles, 1949-1959." *Journal of Urban History* 8, 2 (1982): 123-143.

Nevada

1345. Shepperson, W.S. *Retreat to Nevada; a Socialist Colony of World War I*. Reno, NE: University of Nevada Press, 1966.

Oklahoma

1346. Ameringer, O. *If You Don't Weaken*. New York: Holt and Co., 1940.

1347. Burbank, G. "Agrarian Radicals and Their Opponents: Political Conflict in Southern Oklahoma, 1910-1924." *Journal of American History* 58, 1 (June 1971): 5-23.

1348. ———. "The Disruption and Decline of the Oklahoma Socialist Party." *American Studies* 7, 2 (August 1973): 133-152.

Oregon

1349. Hummasti, P.G. *Finnish Radicals in Astoria, Oregon, 1904-1940: A Study in Immigrant Socialism*. New York: Arno Press, 1979. Reprint.

Washington

1350. Blackford, M.G. "Reform Politics in Seattle during the Progressive Era, 1902-1916." *Pacific Northwest Quarterly* 59, 4 (October 1968): 177-185.

1351. O'Connor, H. *Revolution in Seattle, a Memoir*. New York: Monthly Review Press, 1964.

1352. Schwantes, C.A. *Radical Heritage*. Seattle, WA: University of Washington Press, 1979.

ALL GOD'S PEOPLE IN THE CITY

Religion has served as the symbol of unity for various groups of people throughout history. For many, adverse conditions or a hostile environment have acted to bring about a close adherence to religious principles in order to adapt, accept, and survive. The urban environment is both hostile and yet filled with opportunity for those seeking change or success. Religious groups have always been identifiable within American cities. Randall M. Miller and Thomas D. Marzik's *Immigrants and Religion in Urban America* (1977) is a recent study of the religious identity associated with segments of the urban population. Thomas W. Kremm's "Measuring Religious Preferences in Nineteenth-Century Urban Areas" (*Historic Methods Newsletter* 8, 4 (1975): 137-141) provides a possible methodology for determining the actual religious composition of metropolitan areas. This focus upon religion is reasonable since the connection between religion and other social and political associations in the urban environment is evident from membership lists, voting patterns, and residential choices available from most American urban areas.

Church and Neighborhood (1980) by Edward Kantowicz examines the residential choice of various segments of the Chicago populace. Kantowicz finds that this choice was based upon church affiliation. In Chicago the church was a local neighborhood institution. Irish Catholics lived in St. Gertrude's or St. Lucy's parish. Polish Catholics were concentrated in the area of St. Stanislaus Kostka. These Chicago Catholic parishes defined the political and social life of the surrounding neighborhoods well into the 1960s. Kantowicz demonstrates that religion was the single most important factor in the locus of neighborhood political power. This location of political power within urban religious groups is also noted by Margaret C. Szasz in her article "Albuquerque Congregationalists and Southwestern Social Reform: 1900-1917" (*New Mexico Historical Review* 55, 3 (1980): 231-252). According to Szasz, the First Congregational Church of Albuquerque became identified with social reform in the early twentieth century. Using its pulpit and moral position in the community, the Church sought to rally laborers, immigrants, and the poor within

139

Albuquerque around progressive reform ideas. Education was
the main focus of the Congregationalists' efforts to create
a climate for social reform. Thus, in Chicago and Albuquerque
political power was a feature of local religious institutions
which sought to give direction to the urban life around them.

The following books and journal articles investigate the
influence of religion upon social, political, and economic
actions taken by those who live, work, and play in the urban
environment.

General

1353. Bonkowsky, E.L. *The Church and the City: Protestant
      Concern for Urban Problems 1800-1840*. Boston, MA:
      Boston University Press, 1973.

1354. Davis, D. "A Geographic Perspective of Jewish Atti-
      tudes towards Neighborhood Transition: A Core Study."
      Geographical Perspectives 45 (Spring 1980): 35-48.

1355. Franch, M.S. "The Congregational Community in the
      Changing City, 1840-1870." *Maryland Historical
      Magazine* 71, 3 (Fall 1976): 367-380.

1356. Fried, L. "Jacob Riis and the Jews: The Ambivalent
      Quest for Community." *American Studies* 20, 1
      (Spring 1979): 5-24.

1357. George, W.R. *Segregated Sabbaths: Richard Allen and
      the Rise of Independent Black Churches 1760-1840*.
      New York: Oxford University Press, 1973.

1358. Greene, V. *For God and Country: The Rise of Polish
      and Lithuanian Ethnic Consciousness in America 1860-
      1910*. Madison, WI: State Historical Society of
      Wisconsin, 1975.

1359. Kremm, T.W. "Measuring Religious Preferences in Nine-
      teenth-Century Urban Areas." *Historical Methods
      Newsletter* 8, 4 (September 1975): 137-141.

1360. Kutolowski, K. "Identifying the Religious Affilia-
      tions of Nineteenth-Century Local Elites." *Historical
      Methods Newsletter* 9, 1 (December 1975): 9-13.

1361. Lennon, J.J. *A Comparative Study of the Patterns of
      Acculturation of Selected Puerto Rican Protestant
      and Roman Catholic Families in an Urban Metropolitan
      Area*. San Francisco, CA: R & E Associates, 1977.

1362. Luker, R.E. "Religion and Social Control in the Nine-

teenth-Century American City." *Journal of Urban
History* 2, 3 (May 1976): 363-368.

1363. Marty, M.E. "The Catholic Ghetto and All the Other
Ghettos." *Catholic Historical Review* 68, 2 (April
1982): 184-205.

1364. Miller, R.M., and T.D. Marzik. *Immigrants and Religion in Urban America*. Philadelphia, PA: Temple
University Press, 1977.

1365. Nelson, H.M., and W.E. Snizek. "Musical Pews: Rural
and Urban Models of Occupational and Religious
Mobility." *Sociology and Social Research* 60 (April
1976): 279-289.

1366. Paz, D.G. "The Episcopal Church in Local History Since
1950." *Historical Magazine of the Protestant Episcopal Church* 49, 4 (December 1980): 380-409.

1367. Peters, Victor. "The German Pietists: Spiritual Men-
tors of the German Communal Settlements in America."
Communal Sociology 1 (Autumn 1981): 55-66.

1368. Rischin, M. "Since 1954: A Bicentennial Look at the
Resources of American Jewish History." *The Immigration History Newsletter* 7, 2 (November 1975): 1-6.

1369. Rosenwaike, J. "Estimating Jewish Population Distribution in U.S. Metropolitan Areas in 1970." *Jewish
Social Studies* 36, 2 (April 1974): 106-117.

1370. Rushby, W.F. "The Old German Baptist Brethren: An Intimate Christian Community in Urban-Industrial
Society." *Mennonite Quarterly Review* 21, 4 (October
1977): 362-276.

1371. Singleton, G.H. "Mere Middle-Class Institutions: Urban
Protestantism in Nineteenth-Century America." *Journal
of Social History* 6, 4 (Summer 1973): 489-504.

1372. Sizer, S.S. "Politics and Apolitical Religion: The
Great Urban Revivals of the Late Nineteenth Century."
Church History 48, 1 (March 1979): 81-98.

The East

1373. Akers, C.W. "Religion and the American Revolution:
Samuel Cooper and the Brattle Street Church." *William
and Mary Quarterly* 35, 3 (July 1978): 477-498.

1374. Archer, J. "Puritan Town Planning in New Haven." *Journal*

of the *Society of Architectural Historians* 34 (May
1975): 140-149.

1375. Armstrong, J., and M. Williams. *Building the Mother
Church: The First Church of Christ in Boston, Massa-
chusetts.* Boston, MA: Christian Science Publishing
Society, 1980.

1376. Bardaglio, P.W. "Italian Immigrants and the Catholic
Church in Providence, 1890-1930." *Rhode Island
History* 34, 3 (May 1975): 46-57.

1377. Baughman, E.W. "Excommunications and Banishments from
the First Church in Salem and the Town of Salem 1629-
1680." *Essex Institute Historical Collection* 113, 2
(April 1977): 89-104.

1378. Beck, N.R. "The Use of Library and Educational Facili-
ties by Russian-Jewish Immigrants in New York City,
1880-1914: The Impact of Culture." *Journal of Library
History* 12, 2 (Spring 1977): 129-149.

1379. Benjamen, P.S. *Philadelphia Quakers in the Industrial
Age, 1865-1920.* Philadelphia, PA: Temple University
Press, 1976.

1380. Berenbaum, M. "The Greatest Show That Ever Came to
Town: An Account of the Billy Sunday Crusade in
Buffalo, New York, January 27-March, 1917." *Niagara
Frontier* 22, 3 (Autumn 1975): 54-67.

1381. Berrol, S. "When Uptown Met Downtown: Julia Richardson's Work in the Jewish Community of New York, 1880-
1912." *American Jewish History* 70, 1 (September
1980): 35-51.

1382. Blejwas, S.E. "A Polish Community in Transition: The
Evolution of Holy Cross Parish, New Britain, Con-
necticut." *Polish American Studies* 35, 112 (Spring/
Autumn 1978): 29-44.

1383. Boylan, A.M. "Presbyterians and Sunday Schools in
Philadelphia 1800-1824." *Journal of Presbyterian
History* 58, 4 (Winter 1980): 299-310.

1384. Burnaby, A. "The Jews' Synagogue in Newport, Rhode
Island in 1759." *Rhode Island Jewish Historical
Notes* 7 (November 1975): 33-41.

1385. Curran, R.E. "Prelude to Americanism: The New York
Academia and Clerical Radicalism in the Late Nineteenth Century." *Church History* 47, 1 (March 1978):
48-65.

1386. Delp, R.W. "A Spiritualist in Connecticut: Andrew Jackson Davis, The Hartford Years, 1850-1854." *New England Quarterly* 53, 3 (September 1980): 345-362.

1387. Dolan, J.P. *The Immigrant Churches: New York's Irish and German Catholics 1815-1865.* Baltimore, MD: Johns Hopkins University Press, 1975.

1388. French, R.S. "Liberation from Man and God in Boston: Abner Kneeland's Free Thought Campaign, 1830-1839." *American Quarterly* 32, 2 (Summer 1980): 202-221.

1389. Gaffey, J. "The Changing of the Guard: The Rise of Cardinal O'Connell of Boston." *Catholic Historical Review* 59, 2 (July 1973): 225-244.

1390. Gavigan, K. "The Rise and Fall of Parish Cohesiveness in Philadelphia." *Records of the American Catholic Historical Society, Philadelphia* 86, 114 (March/December 1975): 107-131.

1391. George, J. "Philadelphia's Catholic Herald: The Civil War Years." *Pennsylvania Magazine of History and Biography* 103, 2 (April 1979): 196-221.

1392. Gildrie, R.P. "Contention in Salem: The Hyymson-Nicholet Controversy, 1672-1676." *Essex Institute of Historical Collections* 113, 2 (April 1977): 117-139.

1393. Ginsberg, Y. *Jews in a Changing Neighborhood: The Study of Mattapan.* New York: The Free Press, 1974.

1394. Glasen, R. "The Greek Jews in Baltimore." *Jewish Social Studies* 38, 3/4 (Summer/Fall 1976): 321-336.

1395. Goldberg, A. "The Jew in Norwich, Connecticut: A Century of Jewish Life." *Rhode Island Jewish Historical Notes* 7 (November 1975): 79-103.

1396. Haebler, P. "Holyoke's French-Canadian Community in Turmoil: The Role of the Church in Assimilation, 1869-1887." *Historical Journal of Western Massachusetts* 7 (January 1979): 5-21.

1397. Hershkowitz, L. "Some Aspects of the New York Jewish Merchant and Community, 1654-1820." *American Jewish Historical Quarterly* 66, 1 (September 1976): 10-34.

1398. Hewitt, J.H. "New York's Black Episcopalians: In the Beginning, 1704-1722." *Afro-American New York Life and History* 3 (January 1979): 9-22.

1399. Horvitz, E.F. "Old Bottles, Rags, Funk! The Story of

the Jews of South Providence." *Rhode Island Jewish Historical Notes* 7 (November 1976): 189-257.

1400.   Hyman, P.E. "Immigrant Women and Consumer Protest: The New York City Kosher Meat Boycott of 1902." *American Jewish History* 70, 1 (September 1980): 91-105.

1401.   "Irish Catholics in a Yankee Town: A Report about Brattleboro, Vermont, 1847-1898." *Vermont History* 44, 4 (Fall 1976): 189-197.

1402.   Jacoby, G.P. *Catholic Child Care in Nineteenth-Century New York with a Correlated Summary of Public and Protestant Child Welfare.* New York: Arno Press, 1974. Reprint.

1403.   Jeffries, J.W. "The Separation in the Canterbury Congregational Church: Religion, Family, and Politics in a Connecticut Town." *New England Quarterly* 52, 4 (December 1979): 522-549.

1404.   Johnson, P.E. *A Shopkeeper's Millenium: Society and Revivals in Rochester, New York, 1815-1837.* New York: Hill and Wang, 1978.

1405.   Jones, H.D. *The Evangelical Movement among Italians in New York City: A Study.* New York: Arno Press, 1975. Reprint.

1406.   Juliani, R.N. "Church Records as Social Data: The Italians of Philadelphia in the Nineteenth Century." *Records of the American Catholic Historical Society of Philadelphia* 85, 1/2 (March/June 1974): 3-16.

1407.   Kaplan, M. "The Jewish Merchants of Newport, 1740-1790." *Rhode Island Jewish Historical Notes* 7 (November 1975): 12-32.

1408.   Kring, W.D. *Liberals Among the Orthodox: Unitarian Beginnings in New York City 1819-1839.* Boston, MA: Beacon Press, 1974.

1409.   Lafontaine, C.V. "Sisters in Peril: A Challenge to Protestant Episcopal-Roman Catholic Concord, 1909-1918." *New York History* 58, 4 (October 1977): 440-469.

1410.   Lerner, E. "Jewish Involvement in the New York City Women Suffrage Movement." *American Jewish History* 70, 4 (June 1981): 442-461.

1411.   Lobsen, A.F. "Newport's Jews and the American Revolution." *Rhode Island Jewish Historical Notes* 7 (November 1976): 258-276.

1412. Lurix, K.A. "The German Lutherans in Boston." *Con-cordia Historical Institute Quarterly* 54 (Winter 1981): 146-162.

1413. MacNab, J.B. "Bethlehem Chapel: Presbyterians and Italian Americans in New York City." *Journal of Presbyterian History* 55, 2 (Summer 1977): 145-160.

1414. Mannard, J.G. "The 1839 Baltimore Nunnery Riot: An Episode in Jacksonian Nativism and Social Violence." *The Maryland Historian* 11 (Spring 1980): 13-28.

1415. McElroy, J.L. "Social Control and Romantic Reform in Antebellum America: The Case of Rochester, New York." *New York History* 58, 1 (January 1977): 17-46.

1416. McHale, M.J. *On the Way: The Story of the Pittsburgh Sisters of Mercy, 1834-1968.* New York: Seabury Press, 1980.

1417. Merwick, D. *Boston Priests 1848-1910: A Study of Social and Intellectual Change.* Cambridge, MA: Harvard University Press, 1973.

1418. Moran, G.F. "Religious Renewal, Puritan Tribalism, and the Family in Seventeenth Century, Milford, Connecti-cut." *William and Mary Quarterly* 35, 2 (April 1979): 236-254.

1419. Pierce, R.D. *The Records of the First Church in Salem, Massachusetts: 1629-1736.* Salem, MA: Essex Institute, 1974.

1420. Ralph, R.M. "The City and the Church: Catholic Be-ginnings in Newark 1840-1870." *New Jersey History* 96, 3/4 (Autumn/Winter 1978): 105-119.

1421. Reutilmger, A.S. "Reflections on the Anglo-American Jewish Experience: Immigrants, Workers and Entrepre-neurs in New York and London 1870-1914." *American Jewish Historical Quarterly* 66, 4 (June 1977): 473-484.

1422. Rosenwaike, I. "The Jews of Baltimore: 1810-1820." *American Jewish Historical Quarterly* 67, 3 (March 1979): 246-259.

1423. ———. "The Jews of Baltimore to 1810." *American Jewish Historical Quarterly* 64, 4 (June 1975): 291-321.

1424. Rothchild, S. "Return to Northrup, Massachusetts: Jewish Suburbia, 1975." *Present Tense* 2 (Summer 1975): 36-41.

1425. Ruxin, R.H. "The Jewish Farmer and the Small-Town Jewish Community: Schoharie County, New York." *American Jewish Archives* 29, 1 (April 1977): 3-21.

1426. Ryan, M.P. "A Woman's Awakening: Evangelical Religion and the Families of Utica, New York, 1800-1840." *American Quarterly* 30, 5 (Winter 1978): 602-623.

1427. Ryan, W.A. "The Separation of Church and State in Acworth, New Hampshire." *History of New Hampshire* 34 (Summer 1979): 143-153.

1428. Scheidt, D.L. "The Lutherans in Revolutionary Philadelphia." *Concordia Historical Institute Quarterly* 49 (Winter 1976): 148-159.

1429. Schmandt, R.H. "A Philadelphia Reaction to Pope Pius IX in 1848." *Records of the American Catholic Historical Society Philadelphia* 88, 1/4 (March/December 1977): 63-87.

1430. Schwartz, H. "Adolescence and Revivals in Antebellum Boston." *Journal of Religious History* 8, 2 (December 1974): 144-158.

1431. Selavan, J.C. "Jewish Wage Earners in Pittsburgh, 1890-1930." *American Jewish Historical Quarterly* 65, 3 (March 1976): 272-285.

1432. Sorrell, R.S. "Sentinelle Affair (1924-1929)--Religion and Militant Survivance in Woonset, Rhode Island." *Rhode Island History* 36, 3 (August 1977): 66-79.

1433. Southern, E. "Musical Practices in Black Churches of New York and Philadelphia, ca. 1800-1844." *Afro-American New York Life and History* 4 (January 1980): 61-77.

1434. Sutherland, J.F. "Rabbi Joseph Krauskopf of Philadelphia: The Urban Reformer Returns to the Land." *American Jewish Historical Quarterly* 67, 4 (June 1978): 342-362.

1435. Szarnicki, H.A. *First Catholic Bishop of Pittsburgh, 1843-1860. A Story of the Catholic Pioneers of Pittsburgh and Western Pennsylvania.* Pittsburgh, PA: Wolfson Publishing, 1975.

1436. ————. *Michael O'Connor, First Catholic Pioneer of Pittsburgh and Western Pennsylvania.* Pittsburgh, PA: Wolfson Publishing, 1975.

1437. Tomasi, S.M. *Piety and Power: The Role of Italian Parishes in the New York Metropolitan Area (1880-*

1930). Staten Island, NY: Center for Migration
Studies, 1975.

1438. Williams, R.E. *Called and Chosen: The Story of Mother
Rebecca Jackson and the Philadelphia Shakers.*
Metuchen, NJ: Scarecrow Press, 1980.

1439. Wood, N.E. *The History of the First Baptist Church
of Boston, 1665-1899.* New York: Arno Press, 1980.
Reprint.

1440. Yearwood, L. "First Shiloh Baptist Church of Buffalo,
New York: From Storefront to Major Religious Insti-
tution." *Afro-American New York Life and History* 1
(January 1979): 81-91.

1441. "The Yankee Priest Says Mass in Brattleboro: Joseph
Coolidge Sharo Describes His Visit in 1848." *Vermont
History* 44, 4 (Fall 1976): 198-202.

The Midwest

1442. Achenbaum, W.A. "Towards Pluralism and Assimilation:
The Religious Crisis of Ann Arbor's Wurttenburg
Community." *Michigan History* 58, 3 (Fall 1974):
195-218.

1443. Arndt, K.J. "Luther's Golden Rose at New Harmony,
Indiana." *Concordia Historical Institute Quarterly*
49 (Fall 1976): 112-122.

1444. Bowers, P.C. "Worthington, Ohio: James Kilbourn's
Episcopal Haven on the Western Frontier." *Ohio History* 85, 3 (Summer 1976): 247-262.

1445. Gendler, C. "The First Synagogue in Nebraska: The
Early History of the Congregation of Israel of
Omaha." *Nebraska History* 58, 3 (Fall 1977): 323-
341.

1446. Heath, A.R. "Apostle in Zion (Chicago in the 1890's)."
Journal of the Illinois State Historical Society
70, 1 (May 1977): 98-113.

1447. Jacobs, R.P. *Religion in St. Louis, a Strong Heritage.*
St. Louis, MO: Interfaith Clergy Council, 1976.

1448. Janis, R. "Ethnic Mixture and the Persistence of Cultural Pluralism in the Church Communities of Detroit,
1880-1940." *Mid-America* 61, 2 (April/July 1979):
99-114.

1449. Johnsen, L. "Brounsberger and Battle Creek: The Beginning of Adventist Higher Education." *Adventist Heritage* 3 (Winter 1976): 30-40.

1450. Jones, R.W. "Christian Social Action and the Episcopal Church in Saint Louis, MO: 1880-1920." *Historical Magazine of the Protestant Episcopal Church* 45, 3 (Sept. 1976): 253-279.

1451. Kantowicz, E.R. "Cardinal Mundelein of Chicago and the Shaping of Twentieth-Century American Catholicism." *Journal of American History* 68, 1 (June 1981): 52-68.

1452. ———. "Church and Neighborhood (Chicago)." *Ethnicity* 7, 4 (December 1980): 349-366.

1453. Keefe, T.M. "The Catholic Issue in the Chicago Tribune before the Civil Wars." *Mid-America* 57, 3 (October 1975): 227-245.

1454. Kemper, D.J. "Catholic Integration in St. Louis, 1935-1947." *Missouri Historical Review* 73, 1 (October 1978): 1-22.

1455. Klammer, K.K., et al. "Illinois State University and the Early Lutheran Church of Springfield, Illinois." *Concordia Historical Institute Quarterly* 53 (Winter 1980): 146-165.

1456. Landing, J.E. "Geographic Change among Eastern Christians in the Chicago Area." *Bulletin of the Illinois Geographical Society* 17 (June 1975): 40-47.

1457. Leonard, H.B. "Ethnic Conflict and Episcopal Power: The Diocese of Cleveland 1847-1870." *Catholic Historical Review* 62, 3 (July 1976): 388-407.

1458. McCarthur, B. "1893. The Chicago World's Fair: An Early Test for Adventist Religious Liberty." *Adventist Heritage* 21 (Winter 1975): 11-21.

1459. McGuckin, M. "The Lincoln, Nebraska." *Adventist Heritage* 2 (Summer 1975): 24-31.

1460. Paz, D.G. "A Study in Adoptability: The Episcopal Church in Omaha 1856-1919." *Nebraska History* 62, 1 (Spring 1981): 107-130.

1461. Perko, M. "The Building up of Zion: Religion and Education in Nineteenth-Century Cincinnati." *Cincinnati Historical Society Bulletin* 38, 2 (Summer 1980): 96-114.

1462. Pierce, L.E. "The Jewish Settlement on St. Paul's Lower West Side." *American Jewish Archives* 28, 2 (November 1976): 143-161.

1463. Puotinen, A.E. *Finnish Radicals and Religion in Mid-western Mining Towns 1865-1914*. New York: Arno Press, 1979.

1464. Rahill, P.J. "St. Louis Under Bishop Rosati (1789-1843)." *Missouri Historical Review* 66, 4 (July 1972): 495-519.

1465. ———. "The Genesis of a Communal History: The Columbus Jewish History Project." *American Jewish Archives* 29, 3 (April 1977): 53-69.

1466. ———. "The Utilization of Public Local and Federal Sources for Reconstructing American Jewish Local History: Jews of Columbus, Ohio." *American Jewish Historical Quarterly* 65, 1 (September 1975): 10-35.

1467. Sanders, J.W. *The Education of an Urban Minority: Catholics in Chicago 1833-1965*. New York: Oxford University Press, 1977.

1468. Sievers, A.H. "Judaism in the Heartland: The Jewish Community in Marietta, Ohio (1895-1940)." *Great Lakes Review* 5, 2 (Winter 1979): 24-35.

1469. Stocker, D.S., et al. "The History of the Traverse City Jewish Community." *Michigan Jewish History* 19-20 (June 1979-January 1980): 13-33; 4-19.

1470. Walch, T. "Catholic Education in Chicago: The Formative Years, 1840-1890." *Chicago History* 7, 2 (Summer 1978): 87-97.

1471. ———. "Catholic Social Institutions and Urban Development: The View from Nineteenth-Century Chicago and Milwaukee." *The Catholic Historical Review* 64, 1 (January 1978): 16-32.

1472. Williams, P.W. "The Adams Family in Catholic Cincinnati." *Cincinnati Historical Society Bulletin* 39, 3 (Fall 1981): 195-200.

1473. Young, M.E., and W. Attoe. *Places of Worship in Milwaukee*. Milwaukee, WI: Past-Futures, 1977.

The South

1474. Baker, J.T. "The Battle of Elizabeth City: Christ and Antichrist in North Carolina." *North Carolina Historical Review* 54, 4 (October 1977): 393-408.

1475. Bate, K.W. "Iron City, Mormon Mining Town." *Utah Historical Quarterly* 50, 1 (Winter 1982): 47-58.

1476. Bauman, M.K. "A Famous Atlantan Speaks Out against
       Lynching: Bishop Warren Akin Candler and Social Jus-
       tice." *Atlanta Historical Bulletin* 20 (Spring 1976):
       24-32.

1477. Becker, L.D. "Unitarianism in Postwar Atlanta, 1882-
       1908." *Georgia Historical Quarterly* 56, 3 (Fall
       1972): 349-364.

1478. Carroll, K.L. "The Irish Quaker Community at Camden."
       South Carolina History Magazine 77, 2 (April 1976):
       69-83.

1479. Clarke, T.E. "An Experiment in Paternalism: Presby-
       terians and Slaves in Charleston, South Carolina."
       Journal of Presbyterian History 53, 3 (Fall 1975):
       223-238.

1480. Crane, E.F. "Uneasy Coexistence: Religious Tension
       in Eighteenth-Century Newport." *Newport History* 53
       (Summer 1980): 101-111.

1481. Dayton, W. "Pasco Pioneers: Catholic Settlements in
       San Antonio, St. Leo and Vicinity." *Tampa Bay His-
       tory* 1 (Fall/Winter 1979): 32-39.

1482. Elifson, K.W. "Religious Behavior among Urban Southern
       Baptists: A Casual Inquiry." *Sociological Analysis*
       31, 1 (Spring 1976): 32-45.

1483. Elovitz, M.H. *A Century of Jewish Life in Dixie: The
       Birmingham Experience*. Huntsville, AL: University
       of Alabama Press, 1975.

1484. Flynt, W. "Religion in the Urban South: The Divided
       Religious Mind of Birmingham, 1900-1930." *Alabama
       Review* 30 (April 1977): 108-134.

1485. Flynt, J.W. "Southern Baptists: Rural to Urban Transi-
       tion." *Baptist Historical Heritage* 16 (January
       1981): 24-34.

1486. Hertzberg, S. "Southern Jews and Their Encounter with
       Blacks: Atlanta 1850-1915." *Atlanta Historical
       Journal* 23 (Fall 1979): 7-24.

1487. Holder, R. "Methodist Beginnings in New Orleans, 1813-
       1814." *Louisiana History* 18, 2 (Spring 1977): 171-
       187.

1488. Hoobler, J.A. "Karnak on the Cumberland (Presbyterian-
       ism in Nashville)." *Tennessee Historical Quarterly*
       35, 3 (Fall 1976): 251-262.

1489. Hornbein, M. *Temple Emmanuel of Denver: A Centennial History*. Denver, CO: Congregation Emanuel, 1974.

1490. Johnson, G.M. "Churches and Evangelism in Jackson, Mississippi 1920-1929." *Journal of Mississippi History* 34, 4 (November 1972): 307-329.

1491. Lefever, H.G. "Prostitution, Politics, and Religion: The Crusade against Vice in Atlanta in 1912." *Atlanta Historical Journal* 24 (Spring 1980): 7-29.

1492. Loveland, A.C. "The Southern Work of the Reverend Joseph C. Hartzell, Pastor of Ames Church in New Orleans, 1870-1873." *Louisiana History* 16, 4 (Fall 1975): 391-407.

1493. May, D.L. "Mormon Cooperatives in Paris, Idaho, 1869-1896." *Idaho Yesterdays* 19, 2 (Summer 1975): 20-30.

1494. Newman, H.K. "Piety and Segregation--White Protestant Attitudes toward Blacks in Atlanta, 1865-1905." *Georgia Historical Quarterly* 63, 2 (Summer 1979): 238-257.

1495. O'Brien, J.T. "Factory, Church, and Community: Blacks in Antebellum Richmond." *Journal of Southern History* 44, 4 (November 1978): 509-536.

1496. Parker, H.M. "The Urban Failure of the Southern New School Presbyterian Church." *Social Sciences Journal* 14, 1 (January 1977): 139-148.

1497. Petrusak, F., and S. Skinert. "The Jews of Charleston: Some Old Wine in New Bottles." *Jewish Social Studies* 38, 3/4 (Summer/Fall 1976): 337-345.

1498. Reilly, T. "Genteel Reform Versus Southern Allegiance: Episcopalian Dilemma in Old New Orleans." *Historical Magazine of the Protestant Episcopal Church* 44, 4 (December 1975): 437-450.

1499. Reilly, T.F. "Parson Clapp of New Orleans: Antebellum Social Critic, Religious Radical, and Member of the Establishment." *Louisiana History* 16, 2 (Spring 1975): 167-191.

1500. Rosenwaike, I. "The First Jewish Settlers in Louisville." *Filson Club Historical Quarterly* 53, 1 (January 1979): 37-44.

1501. Sobel, M. "They Can Never Prosper Together: Black and White Baptists in Antebellum Nashville, Tennessee." *Tennesse Historical Quarterly* 38, 3 (Fall 1979): 196-307.

1502.   Surratt, J.L.   "The Role of Dissent in Community Evo-
        lution among Moravians in Salem, 1772-1860." *North
        Carolina Historical Review* 52, 3 (July 1975): 235-
        255.

1503.   Szasz, M.C.   "Albuquerque Congregationalists and South-
        western Social Reform 1900-1917." *New Mexico His-
        torical Review* 55, 3 (July 1980): 231-252.

1504.   Williams, B.S.   "Anti-Semitism in Shreveport, Louisi-
        ana: The Situation in the 1920's." *Louisiana History*
        21, 4 (Fall 1980): 387-398.

1505.   Zweigenhaft, R.L.   "Two Cities in North Carolina: A
        Comparative Study of Jews in the Upper Class."
        Jewish Social Studies 41, 3/4 (Summer/Fall 1979):
        291-300.

The West

1506.   Baker, A.L.   "The San Francisco Evolution Debates:
        June 13-14, 1925." *Adventist Heritage* 2 (Winter
        1975): 23-32.

1507.   Best, G.D.   "Jacob H. Schiff's Galveston Movement: An
        Experiment in Immigrant Deflection, 1907-1914."
        American Jewish Archives 30, 1 (April 1978): 43-79.

1508.   Coogan, M.J.   "Redoubtable John Hennessy, First Arch-
        bishop of Dubuque." *Mid-America* 62, 1 (January 1980):
        21-34.

1509.   Dalin, D.G.   "Jewish and Non-Partisan Republicanism
        in San Francisco, 1911-1963." *American Jewish His-
        tory* 68, 4 (June 1979): 491-516.

1510.   Easton, B., et al.   "Desperate Times: The Peoples
        Temple and the Left." *Socialist Review* 44, 2 (March/
        April 1979): 63-74.

1511.   Kramer, W.M., and N.B. Stern.   "San Francisco's Fight-
        ing Jew." *California Historical Quarterly* 53, 4
        (Winter 1974): 333-346.

1512.   Lamb, B.   "Jews in Early Phoenix 1870-1920." *Journal
        of Arizona History* 18, 3 (Autumn 1977): 299-318.

1513.   May, D.L.   "Mormon Cooperatives in Paris, Idaho, 1869-
        1896." *Idaho Yesterdays* 19, 2 (Summer 1975): 20-30.

1514.   Muller, D.R.   "Church Building and Community Making on
        the Frontier, a Case Study: Josiah Strong, Home

Missionary in Cheyenne, 1871-1873." *Western Historical Quarterly* 10, 2 (April 1979): 191-216.

1515. Senkewicz, R.M. "Religion and Non-Partisan Politics in Gold Rush San Francisco." *Southern California Quarterly* 61 (Winter 1979): 351-378.

1516. Singleton, Gregory H. *Religion in the City of Angels: American Protestant Culture and Urbanization, Los Angeles 1850-1930*. Ann Arbor, MI: UMI Research Press, 1978.

1517. Toll, W. "American Jewish Families: The Occupational Basis of Adaptability in Portland, Oregon." *The Jewish Journal of Sociology* 19, 1 (June 1977): 33-48.

1518. ———. "Fraternalism and Community Structure on the Urban Frontier: The Jews of Portland, Oregon--A Case Study." *Pacific Historical Review* 47, 3 (August 1978): 369-404.

1519. ———. "Mobility, Fraternalism, and Jewish Cultural Change: Portland, 1910-1930." *American Jewish History* 68, 4 (June 1979): 459-491.

1520. ———. "Voluntarism amd Modernization in Portland Jewry: The B'nai Brith in the 1920's." *Western Historical Quarterly* 10, 1 (January 1979): 21-38.

1521. Weber, F.J. "A Catholic Bishop Meets the Racial Problem (Los Angeles, 1920's)." *Records of the American Catholic Historical Society of Philadelphia* 84, 4 (December 1973).

1522. Winn, K. "The Seattle Jewish Community: A Photographic Essay." *Pacific Northwest Quarterly* 70, 2 (April 1979): 69-74.

POLITICS AS USUAL

Distribution of city services and governmental actions
taken by municipal authorities are determined largely by
politics. Indeed, many urban residents maintain that political activities are so intrinsic to events in the urban environment that it is difficult to separate a given event from its
political overtones. The political nature of the municipal
decision-making process is the focus of *Power and Society in
Greater New York, 1886-1903: The Making of Major Decisions in
the Metropolis* (1981) by David C. Hammack.
    A different view of municipal decision-making is presented
by Ira Katznelson's *City Trenches: Urban Politics and the
Patterning of Class in the United States* (1981). Using Marxist
methodology, Katznelson examines the political battles waged
in the Washington Heights-Inwood section of Manhattan during
the 1960s and 1970s. This community was then a multi-ethnic
working class one. During the period studied, the traditional
political power in the community based upon ethnic grouping
was challenged by coalitions of broader-based minority groups
which demanded a city government more responsive to neighborhood needs. Katznelson maintains that this confrontation between traditional ethnic political power groups and minority-
based coalitions dramatically altered the political landscape
not only of New York City but also throughout urban America
which, according to Katznelson, experienced similar challenges.
Type of work as the force for political grouping and action
is, indeed, a Marxist approach to social conditions but ig-
noring the possible bias based upon political orientation of
the reader, Katznelson's work is a significant effort to chart
the dissolution of the traditional governance structure of
many U.S. cities, namely the dominance by ethnic groups in
various city sectors, and the resulting change to a coalition
of minority groups seeking city responsiveness to their needs.
    Jon Teaford's *City and Suburb: The Political Fragmenta-
tion of Metropolitan America* (1979) investigates a different
aspect of city governance: the relationship between the city
and its suburbs. The work provides an analysis of the politics
of metropolitan governmental fragmentation, annexation, and
consolidation. In the 1940s, suburbs achieved an ascendancy

155

over their parent center city. Throughout the 1920s and
1930s, there were attempts to reconcile traditional concepts
of power dispersal with the urge to provide centralized control from the center city over its suburbs by use of the
federal concept: metropolitan government representing all
segments of the connected urban environment. The lack of
cooperation between center city and suburbs prevented a full
federal model creation which brought about the eventual decay
of metropolitan government structure. Suburban independence
from the center city decision-making process followed this
decay. Teaford, however, insists that political fragmentation interferes with the political capacity of both city and
suburb to cope with urgent urban problems.

The following list of books and journal articles is representative of research conducted into the relationship between
politics and urban life.

General

1523. Abney, F.G., and J.D. Hutcheson. "Race, Representation, and Trust: Changes in Attitudes after the Election of a Black Mayor." *Public Opinion Quarterly*
45, 1 (Spring 1981): 91-101.

1524. Allensworth, D.T. *The Political Realities of Urban
Planning*. New York: Praeger, 1975.

1525. Allswang, J.M. *Bosses, Machines and Urban Voters*.
Port Washington, NY: Kennikat Press, 1977.

1526. Bailey, R. *Radicals in Urban Politics: The Alinsky
Approach*. Chicago, IL: University of Chicago Press,
1974.

1527. Bent, D. "Partisan Elections and Public Policy: Response to Black Demands in Large American Cities."
Journal of Black Studies 12 (March 1982): 291-314.

1528. Bernard, R.M., and B.R. Rice. "Political Environment
and the Adoption of Progressive Municipal Reform."
Journal of Urban History 1, 2 (February 1975): 149-
158.

1529. Browning, R.P., and D.R. Marshall. "Minorities and
Urban Electoral Change: A Longitudinal Study." *Urban
Affairs Quarterly* 15, 2 (December 1979): 206-228.

1530. Bryan, F.M. "Does the Town Meeting Offer an Option for
Urban America?" *National Civic Review* 67, 11 (December 1978): 523-528.

1531. Button, J.W. *Black Violence: Political Impact of the 1960's Riots.* Princeton, NJ: Princeton University Press, 1978.

1532. Campbell, A.K., and R.W. Bahl. *State and Local Government: The Political Economy of Reform.* New York: Free Press, 1976.

1533. Caraley, D. "Congressional Politics and Urban Aid: A 1978 Postscript." *Political Science Quarterly* 93, 3 (Fall 1978): 411-420.

1534. Catanese, A.J., and W.P. Farmer. *Personality, Politics, and Planning: How City Planners Work.* Beverly Hills, CA: Sage Publications, 1978.

1535. Cranz, G. *The Politics of Park Design: A History of Urban Parks in America.* Cambridge, MA: MIT Press, 1982.

1536. Danielson, M.N. *The Politics of Exclusion.* New York: Columbia University Press, 1976.

1537. Dorsett, L.W. *Franklin D. Roosevelt and the City Bosses.* Port Washington, NY: Kennikat Press, 1977.

1538. Ebner, M.H. "Urban Government in America 1776-1876." *Journal of American History* 5, 4 (August 1979): 511-520.

1539. ———, and E.M. Tobin. *The Age of Urban Reform: New Perspectives in the Progressive Era.* Port Washington, NY: Kennikat Press, 1977.

1540. Eisinger, P.K. "Black Employment in Municipal Jobs: The Impact of Black Political Power." *American Political Science Review* 76, 2 (1982): 380-382.

1541. ———. *The Politics of Displacement: Racial Ethnic Transition in Three American Cities.* New York: Academic Press, 1980.

1542. Elezar, D.J. *Cities of the Prairie: The Metropolitan Frontier and American Politics.* Lanham, MD: University Press of America, 1984.

1543. Engstrom, R.L., and M.D. McDonald. "The Election of Blacks to City Councils: Clarifying the Impact of Electoral Arrangements on the Sears/Population Relationship." *American Political Science Review* 75, 2 (1981): 344-354.

1544. Exoo, C. "Ethnic Culture and Political Language in Two American Cities." *Journal of Ethnic Studies* 11 (Summer 1983): 79-105.

1545. Fainstein, N.I., and S.S. Fainstein. *Urban Political Movements: The Search for Power by Minority Groups in American Cities*. Englewood Cliffs, NJ: Prentice-Hall, Inc., 1974.

1546. Fisher, C.S. "The City and Political Psychology." *American Political Science Review* 69, 2 (June 1975): 559-571.

1547. Florestano, P.S., and V.L. Marando. "State Commissions on Local Government: Implications for Municipal Officials." *National Civic Review* 67, 8 (1978): 358-362.

1548. Fox, K. *Better City Government: Innovation in American Urban Politics 1850-1937*. Philadelphia: Temple University Press, 1977.

1549. Frisch, M.H. "Urban Theorists, Urban Reform, and American Political Culture in the Progressive Period." *Political Science Quarterly* 97, 2 (1982): 295-316.

1550. Fungiello, P.J. *The Challenge to Urban Liberalism: Federal City Relations during World War II*. Knoxville, TN: University of Tennessee Press, 1978.

1551. Goodman, J.S. *The Dynamics of Urban Government and Politics*. New York: Macmillan, 1975.

1552. Greenstone, D., and P. Peterson. *Race and Authority in Urban Politics: Community Participation and the War on Poverty*. Chicago, IL: University of Chicago Press, 1976.

1553. Griffith, E.S. *A History of American City Government: The Conspicuous Failure 1870-1900*. New York: Praeger, 1974.

1554. Gurr, T.R., et al. *The Politics of Crime and Conflict: A Comparative Study of Four Cities*. Beverly Hills, CA: Sage Publications, 1977.

1555. Hahn, H., and C.H. Levine. *Urban Politics: Past, Present and Future*. New York: Longmans, 1980.

1556. Hammack, D.C. "Elite Perceptions of Power in the Cities of the United States, 1880-1900: The Evidence of James Bryce, Moisei Ostrogorski and Their American Informants." *Journal of Urban History* 4, 4 (August 1978): 363-396.

1557. ———. "Problems in the Historical Study of Power in the Cities and Towns of the United States 1800-1960." *American Historical Review* 83, 2 (April 1978): 323-349.

1558. Hays, S.P. "The Changing Political Structure of the
       City in Industrial America." *Journal of Urban His-
       tory* 1, 1 (November 1974): 6-38.

1559. Hill, R.C. "Separate and Unequal Governmental In-
       equality in the Metropolis." *American Political
       Science Review* 68, 4 (December 1974): 1557-1568.

1560. Hutcheson, J.D., and J.E. Prather. "Economy of Scale
       or Bureaucratic Entropy? Implications for Metro-
       politan Governmental Reorganization." *Urban Affairs
       Quarterly* 15, 2 (1979): 164-182.

1561. Jackson, P.I. "Community Control, Community Mobiliza-
       tion, and Community Political Structure in 57 U.S.
       Cities." *The Sociological Quarterly* 19, 4 (Autumn
       1978): 577-589.

1562. Johnson, A.T. "Electoral Consequences of Neighborhood
       Improvement Projects." *Urban Affairs Quarterly* 16, 1
       (September 1980): 109-116.

1563. Johnson, R.J. "The Political Element in Suburbia: A
       Key Influence on the Urban Geography of the U.S."
       Geography 66, 4 (November 1981): 289-296.

1564. Jones, C.B. "The Impact of Local Election Systems on
       Black Political Representation." *Urban Affairs
       Quarterly* 11, 3 (March 1976): 345-356.

1565. Judd, D.R., and F.N. Kopel. *The Politics of American
       Cities: Private Power and Public Policy.* New York:
       Little, Brown, 1979.

1566. Katznelson, I. *Black Men, White Cities: Race, Politics,
       and Migration in the United States, 1900-68.* Chicago,
       IL: University of Chicago Press, 1976.

1567. Karnig, A.K. "Black Representation in City Councils:
       The Impact of District Elections and Socio-Economic
       Factors." *Urban Affairs Quarterly* 12, 2 (December
       1976): 223-242.

1568. ————. "Local Elections in the U.S.: Separate and
       Unequal." *Current Municipal Problems* 20 (Summer
       1978): 14-19.

1569. ————, and B.O. Walter. "Elections of Women to City
       Councils." *Social Science Quarterly* 56, 1 (March
       1976): 605-613.

1570. Lamb, C. *Political Power in Poor Neighborhoods.* New
       York: Halsted Press, 1975.

1571. Levenstein, C. "The Political Economy of Suburb Civi-
       lizations in Pursuit of a Class Analysis." *The Re-
       view of Radical Political Economics* 13, 2 (Summer
       1981): 23-31.

1572. Levine, C.H. *Racial Conflict and the American Mayor:
       Power, Polarization, and Performance.* Lexington, MA:
       Lexington Books, 1974.

1573. Lewis, E., and F. Anechiarico. *Urban America: Politics
       and Policy.* New York: Holt, Rinehart and Winston,
       1981.

1574. Lewis, E.B. "Attitudes Toward Political Innovation
       among United States City Managers." *The Urban In-
       terest* 2, 1 (Spring 1980): 19-27.

1575. Liebert, R.J. *Disintegration and Political Action:
       The Changing Functions of City Governments in America.*
       New York: Academic Press, 1976.

1576. Long, N.E. "Ethos and the City: The Problem of Local
       Legitimacy." *Ethnicity* 2, 1 (March 1975): 43-52.

1577. McCarthy, M.P. "On Bosses, Reformers, and Urban
       Growth: Some Suggestions for a Political Typology
       of American Cities." *Journal of Urban History* 4, 1
       (November 1977): 29-38.

1578. MacManus, S.A., and C.A. Cassel. "Mexican-Americans
       in City Politics: Participatory Representation and
       Policy Preferences." *Urban Interest* 4 (Spring 1982):
       57-69.

1579. Marando, V.L. "City-County Consolidation: Reform,
       Regionalism, Referenda and Requiem." *Western Politi-
       cal Quarterly* 32, 4 (December 1979): 409-421.

1580. Meltsner, A.J. *The Politics of City Revenue.* Berkeley,
       CA: University of California Press, 1975.

1581. Michalak, T.J., and R.V. Goehlert. *Reform of Local
       Government Structures in the United States 1945-1971.*
       Greenwich, CT: JAI Press, 1976.

1582. Mladenka, K.R. "Citizen Demand and Bureaucratic Re-
       sponse: Direct Dealing Democracy in a Major American
       City." *Urban Affairs Quarterly* 12, 3 (March 1977):
       273-290.

1583. Mollenkopf, J. "Neighborhood and Political Development
       and the Politics of Urban Growth: Boston & San Fran-
       cisco, 1958-1978." *International Journal of Urban
       and Regional Research* 8, 1 (March 1981): 15-39.

1584. Molotch, H. "The City as a Growth Machine: Toward a Political Economy of Place." *American Journal of Sociology* 22, 2 (September 1976): 309-332.

1585. Morlock, L.L., et al. "Civic Elites in Eighty-Eight Cities: Occupational Composition and Its Consequences for Community Decision Making." *South Atlantic Urban Studies*, 3 (1979): 3-37.

1586. Murphy, T.P., and J. Rehfuss. *Urban Politics in the Suburban Era.* Homewood, IL: Dorsey Press, 1976.

1587. Nash, G.B. "The Transformation of Urban Politics." *Journal of American History* 60, 3 (December 1973): 605-632.

1588. ————. *The Urban Crucible: Social Change, Political Consciousness and the Origins of the American Revolution.* Cambridge, MA: Harvard University Press, 1979.

1589. O'Laughlin, J. "The Election of Black Mayors, 1977." *Annals of the Association of American Geographers* 70, 3 (September 1980): 353-370.

1590. Orfield, G. "Federal Policy, Local Power and Metropolitan Segregation." *Political Science Quarterly* 89, 4 (Winter 1974-75): 777-802.

1591. Pienkos, A.T. *Ethnic Politics in Urban America: The Polish Experience in Four Cities.* Chicago: Polish American Historical Association, 1978.

1592. Polsby, N.W. *Community Power and Political Theory: A Further Look at Problems of Evidence and Inference.* New Haven, CT: Yale University Press, 1980.

1593. Potts, J.H. "The Evolution of Municipal Accounting in the United States 1900-1935." *Business History Review* 52, 4 (Winter 1978): 518-536.

1594. Rice, B.R. *Progressive Cities: The Commission Government Movement in America 1901-1920.* Austin, TX: University of Texas Press, 1977.

1595. Richards, P.G. *The Local Government Act of 1972: Problems of Implementation.* Beverly Hills, CA: Sage Publications, 1976.

1596. Robertson, K.A. "Political Impact Analysis in Transportation Planning." *Journal of the Community Development Society of America* 9, 1 (Spring 1978): 112-123.

1597. Salter, J.T. *Boss Rule: Portraits in City Politics.* New York: Arno Press, 1974. Reprint.

1598. Schnall, D.J. *Ethnicity and Suburban Politics*. New
       York: Praeger, 1975.

1599. Schumaker, P.D., and B. Loomis. "Responsiveness to
       Citizen Preferences and Societal Problems in American
       Communities." *South Atlantic Urban Studies* 3 (1979):
       38-66.

1600. Segal, H.P. "Jeff W. Hayes: Reform Boosterism and
       Urban Utopianism." *Oregon Historical Quarterly*
       79, 4 (Winter 1978): 345-358.

1601. Shank, A., and R.W. Conent. *Urban Perspectives: Poli-
       tics and Policies*. Boston: Holbrook Press, 1975.

1602. Steggert, F.X. *Community Action Groups and City Govern-
       ments: Perspectives from Ten American Cities*. Cam-
       bridge, MA: Ballinger Publishing Co., 1975.

1603. Stillman, R.J. *The Rise of the City Manager: A Public
       Professional in Local Government*. Albuquerque, NM:
       University of New Mexico Press, 1979.

1604. Svara, J.H. "Attitudes Toward City Government and
       Preference for District Elections." *South Atlantic
       Urban Studies* 3 (1979): 67-84.

1605. Tabb, W.K., and L. Sawers. *Marxism and the Metropolis:
       New Perspectives in Urban Political Economy*. New
       York: Oxford University Press, 1978.

1606. Teaford, J.C. *City and Suburb: The Political Fragmen-
       tation of Metropolitan Areas, 1950-1970*. Baltimore,
       MD: Johns Hopkins University Press, 1979.

1607. ————. *The Municipal Revolution in America: Origins
       of Modern Urban Government, 1650-1825*. Chicago, IL:
       University of Chicago Press, 1975.

1608. Titus, A.C. "Local Governmental Expenditures and
       Political Attitudes: A Look at Nine Major U.S. Cities."
       Urban Affairs Quarterly 16, 4 (June 1981): 437-452.

1609. Tryman, M.D. "Black Mayoralty Campaigns: Running the
       Race." *Phylon* 35, 4 (December 1974): 346-358.

1610. Vedhtz, A., and E.D. Veblen. "Voting and Contacting:
       Two Forms of Political Participation in a Suburban
       Community." *Urban Affairs Quarterly* 16, 1 (September
       1980): 31-48.

1611. Walton, J. "Community Power and the Retreat from Poli-
       tics; Full Circle after Twenty Years." *Social Prob-
       lems* 23, 3 (February 1976): 293-303.

1612. Young, K. *Essays on the Study of Urban Politics.*
       Hamden, CT: Archon Books, 1975.

1613. Zimmer, T.A. "Urbanization, Social Diversity, Voter
       Turn-Out, and Political Competition in U.S. Elections:
       Analyses of Congressional Districts for 1972." *Social
       Science Quarterly* 56, 1 (March 1976): 689-697.

The East

1614. Anderson, A.D. *The Origin and Resolution of an Urban
       Crisis: Baltimore 1900-1930.* Baltimore, MD: Johns
       Hopkins University Press, 1977.

1615. Arnold, J.L. "The Last of the Good Old Days: Politics
       in Baltimore 1920-1950." *Maryland Historical Maga-
       zine* 71, 3 (Fall 1976): 443-448.

1616. ————. "The Neighborhood and City Hall: The Origin
       of Neighborhood Associations in Baltimore, 1880-
       1911." *Journal of Urban History* 6, 1 (November
       1979): 3-10.

1617. Briceland, A.V. "The Philadelphia Aurora, the New
       England Illuminate, and the Election of 1800."
       Pennsylvania Magazine of History and Biography 100,
       1 (January 1976): 3-36.

1618. Brobeck, S. "Revolutionary Change in Colonial Phila-
       delphia: The Brief Life of the Proprietary Gentry."
       William and Mary Quarterly 33, 3 (July 1976): 410-
       435.

1619. Bukowcyzk, J.J. "The Immigrant Community Reexamined:
       Political and Economic Tensions in a Brooklyn Polish
       Settlement, 1884-1894." *Polish American Studies*
       37, 2 (Autumn 1980): 5-16.

1620. Burkhart, L.C. *Old Values on a New Town: The Politics
       of Race and Class in Columbia, Maryland.* New York:
       Praeger, 1981.

1621. Capeci, D.J. "Fiorello La Guardia and the Stuyvesant
       Town Controversy of 1943." *The New-York Historical
       Society Quarterly* 62, 4 (October 1978): 289-310.

1622. Carey, G.W. "New York: World Economy and Feudal Poli-
       tics." *Focus* 31, 3 (March/April 1981): 1-16.

1623. Caro, R.A. *The Power Broker: Robert Moses and the
       Fall of New York.* New York: Alfred A. Knopf, 1974.

1624. Chalmers, L. "Fiorello La Guardia, Paterfamilias at
       City Hall: An Appraisal." *New York History* 56, 2
       (April 1975): 210-225.

1625. Chern, K.S. "The Politics of Patriotism: War, Ethni-
       city and the New York Mayoral Campaign of 1917."
       New-York Historical Society Quarterly 63, 4 (October
       1979): 291-314.

1626. Cook, E.M. *The Fathers of the Towns: Leadership and
       Community Structures in Eighteenth-Century New
       England.* Baltimore, MD: Johns Hopkins Univ. Press,
       1976.

1627. Cottrol, R.J. "Providence's Black Voters and the
       Dilemma of the 1848 Election." *Southern Studies* 21
       (Fall 1982): 266-276.

1628. Crooks, J.B. "Politics and Reform: The Dimensions of
       Baltimore Progressivism." *Maryland Historical Maga-
       zine* 71, 3 (Fall 1976): 421-427.

1629. Daniels, B.C. "Large Town Office Holding in Eighteenth-
       Century Connecticut: The Growth of Oligarchy." *Jour-
       nal of American Studies* 9, 4 (April 1975): 1-12.

1630. ———. "Town Government in Connecticut, 1636-1675,
       the Founding of Institutions." *The Connecticut Re-
       view* 9, 1 (November 1975): 39-49.

1631. Doherty, R. *Society and Power: Five New England Towns,
       1800-1860.* Amherst, MA: University of Massachusetts
       Press, 1977.

1632. Dubrul, P., and J. Newfield. *The Abuse of Power--The
       Permanent Government and the Fall of New York.* New
       York: Viking Press, 1977.

1633. Ebersole, H.G. "Electricity and Politics in Jamestown,
       New York, 1891-1931." *Niagara Frontier* 22, 1 (Spring
       1975): 22-28.

1634. Ebner, M. "The Future of River City: Passaic, New
       Jersey's Contemporary Urban Political History."
       Urbanism Past and Present 3, 1/2 (Winter 1976/77):
       16-20.

1635. Feinstein, E.F. *Stamford in the Gilded Age, the
       Political Life of a Connecticut Town 1868-1893.*
       Stamford, CT: Stamford Historical Society, 1974.

1636. Fifer, J.V. "Washington, D.C.: The Political Geog-
       raphy of a Federal Capital." *Journal of American
       Studies* 15, 1 (April 1981): 5-26.

1637. Fox, D.M. "Social Policy and City Politics; Tuberculosis Reporting in New York, 1889-1900." *Bulletin of Historical Medicine*, 49 (Summer 1975): 169-195.

1638. Gifford, B.R. "New York City and Cosmopolitan Liberalism." *Political Science Quarterly* 93, 4 (Winter 1978/79): 559-584.

1639. Greene, S.E. "Black Republicans on the Baltimore City Council 1890-1931." *Maryland Historical Magazine* 74, 3 (Fall 1979): 203-222.

1640. Haider, D.H. "Sayre and Kaufman Revisited: New York City Government since 1965." *Urban Affairs Quarterly* 15, 2 (December 1979): 123-145.

1641. Heale, M.J. "From City Fathers to Social Critics: Humanitarianism and Government in New York, 1790-1860." *The Journal of American History* 43, 1 (June 1976): 21-42.

1642. Heckscher, A. *Mayor LaGuardia and New York's Legendary Years*. New York: W.W. Norton, 1978.

1643. Henderson, T.M. *Tammany Hall and the New Immigrants: The Progressive Years*. New York: Arno Press, 1977. Reprint.

1644. Hershkowitz, L. *Tweed's New York: Another Look*. Garden City, NY: Anchor Press, 1977.

1645. Krasner, M. "The Cycle of Powerlessness: School Politics in New York City." *Urban Education* 14, 4 (January 1980): 387-414.

1646. Lefurgy, W.G. "Baltimore's Wards 1797-1978, a Guide." *Maryland Historical Magazine* 75, 2 (June 1980): 145-153.

1647. Lemmey, W. "Boss Kenney of Jersey City 1949-1972." *New Jersey History* 98, 1 (Spring/Summer 1980): 9-29.

1648. Leonard, J.M. "The Politics of Charter Revision in New York City 1845-1847." *The New-York Historical Society Quarterly* 62, 1 (January 1978): 43-70.

1649. Lewinson, E.R. *Black Politics in New York City*. New York: Twayne, 1974.

1650. Marchone, W.P. "The 1949 Boston Charter Reform." *The New England Quarterly* 49, 3 (September 1976): 373-399.

1651. Miller, D.C. *Leadership and Power in the Bos-Wash Megalopolis: Environment, Ecology, and Urban Organization*. New York: John Wiley and Sons, 1975.

1652. Miller, R.G. *Philadelphia the Federalist City: A Study
      of Urban Politics 1789-1801*. Port Washington, NY:
      Kennikat Press, 1976.

1653. Mushkat, J. *The Reconstruction of the New York Demo-
      cracy 1861-1874*. Madison, NJ: Fairleigh Dickinson
      University Press, 1980.

1654. Oaks, R.F. "Philadelphia Merchants and the Origins
      of American Independence." *American Philosophical
      Society Proceedings* 121 (1977): 407-436.

1655. O'Connor, T.H. *Bibles, Brahmins, and Bosses: A Short
      History of Boston*. Boston: Boston Public Library,
      1976.

1656. Pearson, R.L., and L. Wrigley. "Before Mayor Richard
      Lee: George Dudley Seymour and the City Planning
      Movement in New Haven, 1907-1924." *Journal of Urban
      History* 6, 3 (May 1980): 297-319.

1657. Pease, W.H., and J.H. Pease. "Paternal Dilemmas: Edu-
      cation, Property, and Patrician Persistence in Jack-
      sonian Boston." *The New England Quarterly* 53, 2
      (June 1980): 147-167.

1658. Pessen, E. "Who Has Power in the Democratic Capitalist
      Community? Reflections on Antebellum New York City."
      New York History 58, 2 (April 1977): 129-156.

1659. Quandt, J.B. "Community in Urban America, 1890-1917:
      Reformers, City Planners and Greenwich Villagers."
      Societas--A Review of Social History 6, 4 (Autumn
      1976): 255-274.

1660. Ridgway, W.H. "Community Leadership: Baltimore during
      the First and Second Party Systems." *Maryland His-
      torical Magazine* 71, 3 (Fall 1976): 334-348.

1661. ————. *Community Leadership in Maryland, 1790-1840:
      A Comparative Analysis of Power in Society*. Chapel
      Hill, NC: University of North Carolina Press, 1979.

1662. Robinson, F.S. *Machine Politics: A Study of Albany's
      O'Connells*. New Brunswick, NJ: Transaction Books,
      1977.

1663. Rorabaugh, W.J. "Rising Democratic Spirits: Immigrants,
      Temperance, and Tammany Hall, 1854-1860." *Civil War
      History* 22, 2 (June 1976): 138-158.

1664. Ruchelman, L.I. *The World Trade Center: Politics and
      Policies of Skyscraper Development*. Syracuse, NY:
      Syracuse University Press, 1977.

1665. Ryerson, R.A. "Political Movements and the American Revolution: The Resistance Movement in Philadelphia 1756-1776." *William and Mary Quarterly* 31, 4 (October 1974): 565-568.

1666. ———. *The Revolution Is Now Begun: The Radical Committees of Philadelphia 1765-1776.* Philadelphia, PA: Pennsylvania University Press, 1977.

1667. Sauers, B.J. "A Political Process of Urban Growth: Consolidation of the South Side with the City of Pittsburgh, 1872." *Pennsylvania History* 41, 3 (July 1974): 265-287.

1668. Silverman, R.A. "Nathan Matthews: Politics of Reform in Boston, 1890-1910." *The New England Quarterly* 50, 4 (December 1977): 626-643.

1669. Smith, R.M. "The Politics of Pittsburgh Flood Control 1936-1960." *Pennsylvania History* 44, 1 (January 1977): 3-24.

1670. Stewart, I.R. "Politics and the Park: The Fight for Central Park." *The New-York Historical Society Quarterly* 61, 3/4 (July/October 1977): 124-155.

1671. Tobin, E.M. "The Progressive as Humanitarian: Jersey City's Search for Social Justice, 1890-1917." *New Jersey History* 93, 2 (Autumn/Winter 1975): 77-98.

1672. Tomasi, S.M. *Piety and Power: The Role of Italian Parishes in the New York Metropolitan Area (1880-1930).* Staten Island, NY: Center for Migration Studies, 1975.

1673. Trout, C.H. *Boston, The Great Depression and the New Deal.* New York: Oxford University Press, 1977.

1674. Weinbaum, P.O. "Temperance, Politics, and the New York City Riots of 1857." *The New-York Historical Society Quarterly* 59, 3 (July 1975): 246-270.

1675. Weinberg, L.S. "Stability and Change among Pittsburgh Precinct Politicians, 1954-1970." *Social Science* 50, 1 (Winter 1975): 10-17.

1676. Weinstein, B.L. "The Demographics and Politics of Economic Decline in New York City." *Annals of Regional Science* 11, 2 (July 1977): 65-73.

The Midwest

1677. Buekner, J.D. "Dynamics of Chicago Ethnic Politics
       1900-1930." *Journal of the Illinois State Historical
       Society* 67, 1 (April 1974): 175-199.

1678. Busch, R.J., and M.D. Abravanel. "The Urban Party
       Organization as an Opportunity Structure: Race and
       Party Differences among Developed Ward Leaders."
       Western Political Quarterly 29, 1 (March 1976):
       7-28.

1679. Cary, L.L. "The Bureau of Investigation and Radicalism
       in Toledo, Ohio: 1918-1920." *Labor History* 21, 3
       (Summer 1980): 430-440.

1680. Chatterjee, P. *Local Leadership in Black Communities:
       Organizational and Electoral Development in Cleveland
       in the Nineteen Sixties*. Cleveland, OH: Case Western
       Reserve University, 1975.

1681. Elenbaas, J.D. "The Boss of the Better Class: Henry
       Leland and the Detroit Citizen's League 1912-1924."
       Michigan History 58, 2 (Summer 1974): 131-151.

1682. Ettenheim, S.C. *How Milwaukee Voted*. Milwaukee, WI:
       University of Wisconsin-Milwaukee Press, 1980.

1683. Ewen, L.A. *Corporate Power and Urban Crisis in De-
       troit*. Princeton, NJ: Princeton University Press,
       1978.

1684. Fragnoli, R.R. "Progressive Coalition and Municipal
       Reform: Charter Revision in Detroit 1912-1918."
       Detroit in Perspective 4 (Spring 1980): 119-142.

1685. Frank, C.M. "Who Governed Middletown? Community
       Power in Muncie, Indiana in the 1930's." *Indiana
       Magazine of History* 75, 4 (December 1979): 321-343.

1686. Funchion, M.F. "Irish Nationalists and Chicago Poli-
       tics in the 1800's." *Erie-Ireland* 10 (Summer 1975):
       3-19.

1687. Giffin, W.W. "The Political Realignment of Black
       Voters in Indianapolis, 1924." *Indiana Magazine of
       History* 79 (June 1983): 133-166.

1688. Goist, P.D. "Political Campaigning and Pluralism in
       a Cleveland Suburb." *Indiana Social Studies Quarterly*
       28, 2 (Autumn 1975): 103-113.

1689. Guterbock, T.M. *Machine Politics in Transition: Party
       and Community in Chicago*. Chicago, IL: University of
       Chicago Press, 1981.

1690. Harrigan, J.T., and W.C. Johnson. *Governing the Twin Cities Region: The Metropolitan Council in Comparative Perspective.* Minneapolis, MN: University of Minnesota Press, 1978.

1691. Hertz, S.H. "The Institutional Context and Political Organization among Welfare Recipients in Minneapolis." *South Atlantic Urban Studies* 2 (1978): 123-140.

1692. Huch, R.K. "Typhoid, Truelson, Water, and Politics in Duluth, 1886-1900." *Minnesota History* 476, 1 (Spring 1981): 189-199.

1693. Hunter, D.A. "The Aftermath of Carl Stokes: An Analysis of Political Drama in the 1971 Cleveland Mayoral Campaign." *Journal of Black Studies* 8, 3 (March 1978): 337-354.

1694. Johnson, R.M. "Politics and Pedagogy: The 1982 Cleveland School Reform." *Ohio History* 84, 4 (Autumn 1975): 196-206.

1695. Johnson, W., and J.J. Harrigan. "Innovation by Increment: The Twin Cities as a Case Study in Metropolitan Reform." *Western Political Quarterly* 31, 2 (June 1978): 206-218.

1696. Jones, G.D. "The Origin of the Alliance between the New Deal and the Chicago Machine." *Journal of the Illinois State Historical Society* 67, 2 (June 1974): 253-271.

1697. Judis, J. "Decline and Fall: Cleveland Says No to Kucinich's Experiment in Urban Populism." *Progressive* 44 (January 1980): 36-39.

1698. Kamphoefner, W.D. "St. Louis Germans and the Republican Party 1848-1860." *Mid-America* 57, 2 (April 1975).

1699. Kantowicz, E.R. *Polish American Politics in Chicago 1880-1940.* Chicago, IL: Chicago University Press, 1976.

1700. Kilian, M. *Who Runs Chicago?* New York: St. Martin's Press, 1979.

1701. Klement, F.L. "Sound and Fury: Civil War Dissent in the Cincinnati Area." *Cincinnati Historical Society Bulletin* 35, 2 (Summer 1977): 99-114.

1702. Lineberry, R.L., and S.M. Watson. "Neighborhoods, Politics, and Public Services: The Case of Chicago." *Urban Interest* 2, 1 (Spring 1980): 11-18.

1703. Longstreet, S. *Chicago: An Intimate Portrait of
       People, Pleasures and Power 1860-1919.* New York:
       David McKay Company, 1973.

1704. Lorenz, A.L. "Rufus King: Politics and Journalism in
       Early Milwaukee." *Historical Messenger of the Mil-
       waukee County Historical Society* 33, 3 (Autumn 1977):
       66-76.

1705. Madison, J.H. "Business and Politics in Indianapolis:
       The Branch Bank and the Junto, 1837-1846." *Indiana
       Magazine of History* 71, 1 (March 1975): 1-21.

1706. Marcus, A.I. "Professional Revolution and Reform in
       the Progressive Era: Cincinnati Physicians and the
       City Elections of 1887 and 1900." *Journal of Urban
       History* 5, 2 (February 1979): 183-208.

1707. Oster, D.B. "Reformers Factionalists, and Kansas City's
       1925 City Manager." *Missouri Historical Review* 72, 3
       (April 1978): 296-327.

1708. Peterson, P.E. *School Politics, Chicago Style.* Chicago,
       IL: University of Chicago Press, 1977.

1709. Petrocik, J.R. "Voting in a Machine City: Chicago
       1975." *Ethnicity* 8, 3 (September 1981): 320-340.

1710. Proress, D.L. "Banfield's Chicago Revisited: The Con-
       ditions for and Social Policy Implications of the
       Transformation of a Political Machine." *Social
       Service Review* 48, 2 (June 1974): 184-202.

1711. Rakove, M. *Don't Make No Waves ... Don't Back No
       Losers: An Insider's Account of the Daley Machine.*
       Bloomington, IN: Indiana University Press, 1975.

1712. Ranney, J.A. "The Political Campaigns of Mayor David S.
       Rose." *Milwaukee History* 4, 1 (Spring 1981): 2-19.

1713. Reichard, M. "Urban Politics in Jacksonian St. Louis:
       Traditional Values in Change and Conflict." *Missouri
       Historical Review* 70, 3 (April 1976): 259-271.

1714. Reynolds, M.S. "The City, Suburbs, and the Establish-
       ment of the Clifton Town Meeting 1961-1964." *Cincin-
       nati Historical Society Bulletin* 38 (Spring 1980):
       6-32.

1715. Roberts, D.J. "A Chicago Political Diary, 1928-1929."
       Journal of the Illinois State Historical Society
       71, 1 (February 1978): 30-56.

1716. Richardson, J.R. "Urban Political Change in the Pro-
       gressive Era." *Ohio History* 87, 3 (Summer 1978): 310-
       321.

1717. Salces, Luis. "Spanish-Americans Search for Political Representation: The 1975 Aldermanic Election in Chicago." *Journal of Political and Military Sociology* 6, 2 (Fall 1978): 175-188.

1718. Salem, G. *The Forty-Fourth Ward Assembly: An Experiment in Neighborhood Democracy.* Chicago, IL: Loyola University of Chicago, 1980.

1719. Steiner, F. *The Politics of New Town Planning: The Newfields, Ohio Story.* Athens, OH: University of Ohio Press, 1981.

1720. Williams, B. "Petticoats in Politics: Cincinnati Women and the 1920 Election." *Cincinnati Historical Society Bulletin* 35, 1 (Spring 1977): 43-70.

1721. Williams, L. "Newcomers to the City: A Study of Black Populism Growth in Toledo, Ohio, 1910-1930." *Ohio History* 89, 1 (Winter 1980): 5-24.

1722. Woody, C.H. *The Chicago Primary of 1926: A Study in Election Methods.* New York: Arno Press, 1974. Reprint.

The South

1723. Brescia, A.M. "The Election of 1828: A View from Louisville." *Register of the Kentucky Historical Society* 74, 1 (January 1976): 51-57.

1724. Burman, S. "The Illusion of Progress: Race and Politics in Atlanta, Georgia." *Ethnic and Racial Studies* 2, 4 (October 1979): 441-454.

1725. Cotrell, C.L., and R.M. Stevens. "The 1975 Voting Rights Act and San Antonio, Texas: Toward a Federal Guarantee of a Republican Form of Local Government." *Publius* 8, 1 (Winter 1978): 79-100.

1726. Fink, L. "Irrespective of Party, Color or Social Standing: The Knights of Labor and Opposition Politics in Richmond, Virginia." *Labor History* 19, 3 (Summer 1978): 325-349.

1727. Freyer, T.A. "Politics and Law in the Little Rock Crisis, 1954-1957." *Arkansas Historical Quarterly* 40, 1 (Autumn 1981): 195-219.

1728. Haas, E.F. *De Lesseps S. Morrison and the Image of Reform: New Orleans Politics, 1946-1961.* Baton Rouge, LA: Louisiana State University Press, 1974.

1729. ———. "John Fitzpatrick and Political Continuity
       in New Orleans, 1896-1899." *The Journal of the
       Louisiana Historical Association and the Louisiana
       Historical Society* 22, 1 (Winter 1981): 7-30.

1730. Harris, C.V. *Political Power in Birmingham, Alabama
       1871-1921.* Knoxville, TN: University of Tennessee
       Press, 1977.

1731. Henderson, W.D. *Gilded Age Politics: Life and Labor
       in Petersburg, Virginia, 1874-1889.* Lanham, MD:
       University Press of America, 1980.

1732. Hunter, F. *Community Power Succession: Atlanta's
       Policy-Makers Revisited.* Chapel Hill, NC: University
       of North Carolina Press, 1980.

1733. Issac, P.I. "Municipal Reform in Beaumont, Texas,
       1902-1909." *Southwestern Historical Quarterly* 78, 4
       (April 1975): 409-430.

1734. Johnson, M.P. "Planters and Patriarchy: Charleston,
       1800-1860." *Journal of Southern History* 46, 1
       (February 1980): 45-72.

1735. Jordan, L.W. "Police and Politics: Charleston in the
       Gilded Age 1880-1900." *South Carolina Historical
       Magazine* 81, 1 (January 1980): 35-50.

1736. Lyons, W.E. *The Politics of City-County Merger: The
       Lexington-Fayette County Experience.* Lexington, KY:
       University Press of Kentucky, 1977.

1737. Nassbaum, R.O. "The Ring Is Smashed: The New Orleans
       Municipal Election of 1896." *Louisiana History*
       17, 3 (Summer 1976): 283-299.

1738. Pease, J.H., and W.H. Pease. "The Economics and Poli-
       tics of Charleston's Nullification Crisis." *The
       Journal of Southern History* 47, 3 (August 1981):
       335-362.

1739. Ranken, D.C. "The Origins of Black Leadership in New
       Orleans during Reconstruction." *Journal of Southern
       History* 40, 3 (August 1974): 417-440.

1740. Schott, M.L.J. "The New Orleans Machine and Progressi-
       vism." *Louisiana History* 24 (Spring 1983): 141-153.

1741. Tucker, D.M. *Memphis since Crump: Bossism, Blacks and
       Civic Reformers 1948-1968.* Knoxville, TN: University
       of Tennessee Press, 1979.

1742. Wald, K.D. "The Electoral Base of Political Machines:

A Deviant Case Analysis (Memphis)." *Urban Affairs Quarterly* 16, 1 (September 1980): 3-30.

1743. Watts, E.J. "Black Political Progress in Atlanta 1868-1895." *Journal of Negro History* 59, 3 (July 1974): 268-286.

1744. ———. "Property and Politics in Atlanta, 1865-1903." *Journal of Urban History* 3, 3 (May 1977): 295-322.

1745. ———. *The Social Basis of City Politics Atlanta 1865-1903.* Westport, CT: Greenwood Press, 1978.

1746. Woods, S. *Mexican Ethnic Leadership in San Antonio, Texas.* New York: Arno, 1976. Reprint.

The West

1747. Blackford, M.G. "Civic Groups, Political Action, and City Planning in Seattle 1892-1915." *Pacific Historical Review* 49, 4 (November 1980): 557-580.

1748. Bullough, W.A. *The Blind Boss and His City: Christopher Augustine Buckley and Nineteenth-Century San Francisco.* Berkeley, CA: University of California Press, 1979.

1749. ———. "Hannibal versus the Blind Boss: The Junta, Chris Buckley, and Democratic Reform Politics in San Francisco." *Pacific Historical Review* 46, 2 (May 1977): 181-206.

1750. Clements, K.A. "Politics and the Park: San Francisco's Fight for Hetch Hetchy 1908-1913." *Pacific Historical Review* 48, 2 (May 1979): 189-216.

1751. Dalin, D.G. "Jewish and Non-Partisan Republicanism in San Francisco 1911-1963." *American Jewish History* 68, 4 (June 1979): 491-516.

1752. D'Emilio, J. "Gay Politics, Gay Community: San Francisco's Experience." *California History* 59, 1 (1980/81): 334-351.

1753. Hansen, K.T. "Ethnic Group Policy and the Politics of Sex: The Seattle Indian Case." *Urban Anthropology* 8, 1 (Spring 1979): 29-48.

1754. Issel, W. "Class and Ethnic Conflict in San Francisco Political History: The Reform Charter of 1898." *Labor History* 18, 3 (Summer 1977).

1755. Johnson, D.R. "Equal Rights and the Heathen Chinee:
       Black Activism in San Francisco." *Western Historical
       Quarterly* 11, 1 (January 1980): 57-68.

1756. Kahn, J. *Imperial San Francisco: Politics and Planning
       in an American City 1897-1906*. Lincoln, NE: Uni-
       versity of Nebraska Press, 1979.

1757. Lee, E.C., and J. Rothman. "San Francisco's District
       System Alters Electoral Politics." *National Civic
       Review* 67, 4 (April 1978): 173-178.

1758. Lewin, D. "Local Government Labor Relations in Transi-
       tion: The Case of Los Angeles." *Labor History* 17, 2
       (Spring 1976): 191-213.

1759. Lotchin, R.W. "The City and the Sword: San Francisco
       and the Rise of the Metropolitan-Military Complex
       1919-1941." *Journal of American History* 65, 4 (March
       1979): 996-1020.

1760. ————. "The Darwinian City: Politics of Urbanization
       in San Francisco between the World Wars." *Pacific
       Historical Review* 48, 3 (August 1979): 357-383.

1761. MacColl, E.K. *The Growth of a City: Power and Politics
       in Portland Oregon: 1915-1950*. Portland: Georgian
       Press, 1979.

1762. ————. *The Shaping of a City: Business and Politics
       in Portland, Oregon 1885-1915*. Portland: Georgian
       Press, 1976.

1763. Pendergrass, L.F. "The Formation of a Municipal Re-
       form Movement: The Municipal League of Seattle."
       Pacific Northwest Quarterly 66, 1 (January 1975):
       13-26.

1764. Pressman, J.L. *Federal Programs and City Politics:
       The Dynamics of the Aid Process in Oakland*. Ber-
       keley, CA: University of California Press, 1978.

1765. Shearer, D. "How the Progressives Won in Santa Monica."
       Social Policy 12, 3 (Winter 1982): 7-14.

1766. Schiesl, E.M. "Progressive Reform in Los Angeles under
       Mayor Alexander, 1909-1913." *California Historical
       Quarterly* 54, 1 (Spring 1975): 37-56.

1767. Suber, H. "Politics and Popular Culture: Hollywood
       at Bay, 1933-195-." *American Jewish History* 68, 4
       (June 1979): 517-533.

1768. Thompson, F.J. *Personnel Policy in the City: The Politics of Jobs in Oakland*. Berkeley, CA: University of California Press, 1975.

1769. Wirt, F.M. *Power in the City: Decision Making in San Francisco*. Berkeley, CA: University of California Press, 1975.

SCHOOLING THE CITY

An informed and educated electorate has often been identified as the essential requirement for an effective democratic form of government. Since the late eighteenth century
when the Northwest Ordinance became law, the United States
has acted to insure an education for the majority, if not
all, of its citizens. Urban residents have benefited greatly
from this feature of American democracy even from the early
years of the Republic. Ronald Cohen's "Schooling in Early
Nineteenth-Century Boston and New York" (*Journal of Urban
History* 1, 1 (1974): 116-123) is an interesting review essay
concerning the creation and maintenance of schools in two of
the nation's leading cities in the early years of urban growth.
A more detailed analysis of schools in a number of cities
throughout the century following the creation of the United
States is found in David A. Reeder's *Urban Education in the
Nineteenth Century* (1978).

Present research into the creation of schools in American
cities during the formative years of the nation is well represented by David L. Angus's "Detroit's Great School Wars:
Religion and Politics in a Frontier City 1842-1853" (*Michigan
Academician* 12, 3 (1980): 261-280). This article provides an
early view of the still divisive issue of using public tax
money to support sectarian schools. In 1853, Detroit's mayoral
election was waged over this issue according to Angus. The
analysis of the ethnic and religious composition of the wards
in the city of Detroit is an important aspect of Angus's
work. Schools and financial support for education became the
focus for the 1853 election in Detroit. An electoral loss for
the faction supporting the use of public taxes for private
schools did not end the issue but this loss provides an interesting glimpse of the emotions and political actions brought
to bear upon the methods by which education was provided in
the urban setting.

Elections and schools are common subjects for several recent works concerned with urban education. Anne E. Just's
"Urban School Board Elections: Changes in the Political Environment Between 1950 and 1980" (*Education and Urban Society*
12, 4 (1980): 421-435) examines the efforts of many since the

1890s to purge the excesses of partisan politics from the
administration of urban schools. This reform process continued
well into the 1950s and diminished the influence political
parties had on the recruitment and selection of candidates
for school boards. However, Just notes that there is evidence
from the 1970s showing there has been a return to partisan
support for school-board candidates. This support is no
longer political party in orientation but interest-group
oriented. School-board elections in New York City are the
focus of Michael Krasner's "The Cycle of Powerlessness:
School Politics in New York City" (*Urban Education* 14, 4
(1980): 388-414). Krasner details the changes in school
board elections following the passage of the Decentralization
Act of 1969. This law created elected community/neighborhood
school boards which had some authority over elementary schools
but were ultimately controlled by the central school board.
Krasner concludes from his research that during the first
five years of the law's existence, there was a bias evident
in school politics against the poor and minority groups.

The following list of books and journal articles provides
a sampling of recent research into the school as an urban institution.

General

1770. Berrol, S.C. "Julia Richman: Agent of Change in the
      Urban School." *Urban Education* 11, 4 (January 1977):
      357-374.

1771. ———. "Urban Schools: The Historian as Critic."
      Journal of Urban History 8, 2 (February 1982): 206-
      216.

1772. Berube, M.R. *The Urban University in America*. West-
      port, CT: Greenwood Press, 1978.

1773. Black, W.L. "Education in the South from 1820 to 1860
      with Emphasis on the Growth of Teacher Education."
      Louisiana Studies 12, 1 (Winter 1973): 617-629.

1774. Bone, R.G. "Free Public Education: 1825-1860." *Illi-
      nois History* 31 (December 1977): 51-53.

1775. Boyd, W.L. "Educational Policy Making in Declining
      Suburban School Districts: Some Preliminary Findings."
      Education and Urban Society 11, 3 (1979): 333-366.

1776. Brown, F. "Problems and Promises of Urban Public
      Schools." *Journal of Negro Education* 44, 3 (Summer
      1975): 247-256.

1777. Bullock, C.S., and H. Rodgers. "Coercion to Compliance: Southern School Districts and School Desegregation Guidelines." *Journal of Political Science* 38, 2 (November 1976): 987-1011.

1778. Butts, R.F. "Public Education and Political Community." *History of Education Quarterly* 14, 2 (Summer 1975): 165-183.

1779. Carper, J.C., and C.E. Litz. "Reflections on Traditionalist and Revisionist Interpretations of American Educational History." *High School Journal* 61, 2 (December 1977): 119-121.

1780. Cataldo, E.F., et al. "Metropolitan School Desegregation: Practical Remedy or Impractical Ideal?" *Annals of the American Academy of Political and Social Science* 422 (November 1975): 97-104.

1781. Cohen, R.D. "Schooling in Early Nineteenth-Century Boston and New York." *Journal of Urban History* 1, 1 (November 1974): 116-123.

1782. Coleman, J.S. "School Desegregation in Large Cities: A Critique of the Coleman White Flight Thesis." *Harvard Education Review* 46, 2 (May 1976): 217-224.

1783. Davidoff, S.H. "The Crossroad of Urban Education: Title I E.S.E.A." *Urban Education* 9, 2 (July 1974): 152-160.

1784. Eaton, W.E. "Jesse Jackson and the Urban School Reform Movement." *Urban Education* 11, 4 (January 1977): 397-402.

1785. Edson, C.H. "Sociocultural Perspectives on Work and Schooling in Urban America." *Urban Review* 11, 3 (Fall 1979): 127-148.

1786. Finkelstein, B. "In Fear of Childhood: Relationships between Parents and Teachers in Popular Primary Schools in the Nineteenth Century." *History of Childhood Quarterly* 3, 3 (Winter 1976): 321-336.

1787. Freeman, R.B. "Political Power, Desegregation, and Employment of Black School Teachers." *Journal of Political Economy* 85, 2 (April 1977): 299-322.

1788. Glazer, N. "Ethnicity and the Schools." *Commentary* 58, 9 (September 1974): 55-59.

1789. Graf, H.J. *The Literacy Myth: Literacy and Social Structure in the Nineteenth-Century City.* New York: Academic Press, 1979.

1790. Guilmet, G.M. "Material Perceptions of Urban Navajo
      and Caucasian Children's Classroom Behavior."
      Human Organization 38, 1 (1979): 87-91.

1791. Henkin, A.B., and C.D. Ignasias. "Planning in Public
      Education: Lessons from Horace Mann and Henry Bar-
      nard." *Clearing House* 51, 7 (May 1978): 430-435.

1792. Hudgins, H.C. "Brown and Public Segregation: 25 Years
      Ago." *NOLPE School Law Journal* 8, 2 (1979).

1793. Hunt, T.C. "Public Schools, Americanism, and the Im-
      migrant at the Turn of the Century." *Journal of
      General Education* 26, 2 (Summer 1974): 147-155.

1794. Just, A.E. "Urban School Board Elections: Changes
      in the Political Environment between 1950 and 1980."
      Education and Urban Society 12, 4 (August 1980):
      421-435.

1795. Katz, M.B. "Origins of Public Education: A Reassess-
      ment." *History of Education Quarterly* 16, 4 (Winter
      1976): 381-407.

1796. Katznelson, I., et al. "Public Schooling and Working-
      Class Formation: The Case of the United States (1870-
      1900)." *American Journal of Education* 90, 2 (Feb-
      ruary 1982): 111-143.

1797. McClellan, B.E. "Moral Education and Public Schooling:
      An Historical Perspective." *Viewpoints* 51, 6 (No-
      vember 1975): 1-15.

1798. Mohl, R.A. "Urban Education in the Twentieth Century:
      Alice Barrows and the Platoon School Plan." *Urban
      Education* 9, 3 (October 1974): 213-237.

1799. Mohraz, J.J. *The Separate Problem: Case Studies of
      Black Education in the North, 1900-1930.* Westport,
      CT: Greenwood Press, 1979.

1800. Morris, D.J.S. "Public Schooling in the United States."
      Journal of American Studies 9, 2 (August 1975): 229-
      238.

1801. Parkay, F. "Inner City High School Teachers: The Re-
      lationship of Personality Traits and Teaching Style
      to Environmental Stress." *Urban Education* 14, 4
      (January 1980): 449-470.

1802. Raggatt, P. "Equality of Educational Opportunity for
      Minority Group Students: Participants and Policies
      1945-1971." *Comparative Education* 12 (March 1976):
      45-53.

1803.   Reed, D.B., and D.A. Corners. "The Vice Principalship
        in Urban High Schools: A Field Study." *Urban Edu-
        cation* 16, 4 (January 1982): 465-482.

1804.   Reeder, D.A. *Urban Education in the Nineteenth Cen-
        tury.* New York: St. Martin's Press, 1978.

1805.   Reese, W.J. "The Control of Urban School Boards during
        the Progressive Era: A Reconstruction." *Pacific
        Northwest Quarterly* 68, 4 (October 1977): 164-179.

1806.   Rodgers, H.R. "The Supreme Court and School Desegrega-
        tion: Twenty Years Later." *Political Science Quar-
        terly* 89, 4 (Winter 1974-75): 751-776.

1807.   Roper, D. "Coming Full Circle; Charity Schools in
        Public Education." *Social Studies* 71, 2 (March/April
        1980): 90-94.

1808.   Rossell, C.H. "School Desegregation and White Flight."
        Political Science Quarterly 90, 4 (Winter 1975-76):
        675-695.

1809.   Rothstein, S.W. "Orientations: First Impressions in
        the Urban Junior High School." *Urban Education*
        14, 1 (January 1979): 91-116.

1810.   Rubinowitz, H.N. "Half a Loaf: The Shift from White
        to Black Teachers in the Negro Schools of the Urban
        South 1865 to 1890." *Journal of Social History*
        40, 1 (Nov. 1974): 565-594.

1811.   Russo, Frances X. "John Howland: Pioneer in the Free
        School Movement." *Rhode Island History* 37, 4
        (November 1978): 111-122.

1812.   Simpson, R.J. "Brown I: The Historical Perspective."
        NOLPE School Law Journal 8, 2 (1979): 107-115.

1813.   Small, S.E. *The Yankee Schoolmarm in Southern Freed-
        man's Schools, 1861-1871: The Career of a Stereotype.*
        Pullman, WA: Washington State University Press,
        1976.

1814.   Smith, M.P. "Elite Theory and Policy Analysis: The
        Politics of Education in Suburbia." *Journal of
        Political Science* 36, 2 (November 1974): 1006-1032.

1815.   Swanson, A.D., and R.A. King. "The Impact of the
        Courts on the Financing of Public Schools in Large
        Cities." *Urban Education* 11, 2 (July 1976): 15-66.

1816.   Thomas, W.B. "Urban Schooling for Black Migrant Youth:
        A Historical Perspective, 1915-1925." *Urban Educa-
        tion* 14, 3 (October 1979): 267-284.

1817. Tobin, G.A., and B.D. Anderson. "Will Public Schools
       Benefit from Urban Redevelopment?" *Urban Education*
       17 (April 1982): 73-96.

1818. Tyack, D.B. "Pilgrim's Progress: Toward a Social His-
       tory of the School Superintendency, 1860-1960."
       History of Education Quarterly 16, 3 (Fall 1976):
       257-300.

1819. Tyack, D.B., and E. Hanslot. "From Social Movement
       to Professional Management: An Inquiry into the
       Changing Character of Leadership in Public Education."
       American Journal of Education 88, 3 (May 1980): 291-
       319.

1820. Urban, W. "Organized Teachers and Educational Reform
       during the Progressive Era: 1890-1920." *History of
       Education Quarterly* 16, 1 (Spring 1976): 35-52.

1821. Vaughan, S. "To Create a Nation of Noble Men: Public
       Education, National Unity, and the National School
       Service, 1918-1919." *Historian* 41, 3 (May 1979):
       429-449.

1822. Violas, P.C. *The Training of the Urban Working Class:
       A History of Twentieth-Century American Education.*
       Chicago, IL: Rand McNally, 1978.

The State and Public Schools

1823. Abney, E.E. "The Status of Florida's Black School
       Principals." *Journal of Negro Education* 43, 1
       (Winter 1974): 3-8.

1824. Belding, R.E. "An Iowa School Girl--1860's Style."
       Palimpsest 58, 1 (January/February 1977): 2-11.

1825. Carper, J.P. "The Popular Ideology of Segregated
       Schooling: Attitudes Toward the Education of Blacks
       in Kansas, 1854-1900." *Kansas History* 1, 4 (Winter
       1978): 254-265.

1826. Cain, L.C. "Founding Public Schools in Alabama--A
       County Led the Way." *Alabama Historical Quarterly*
       38, 4 (Winter 1976): 243-249.

1827. DeLeon, A. "Blowout 1910 Style: A Chicano School Boy-
       cott in West Texas." *Texana* 12, 2 (1974): 124-140.

1828. Field, A.J. "Economic and Demographic Determinants of
       Educational Commitment: Massachusetts, 1855." *Journal
       of Economic History* 39, 2 (June 1979): 439-459.

1829. ————. "Educational Expansion in Mid-Nineteenth-
       Century Massachusetts: Human Capital Formation or
       Structural Reinforcement." *Harvard Education Review*
       46, 4 (November 1976): 521-552.

1830. ————. "Educational Reform and Manufacturing Develop-
       ment in Mid-Nineteenth-Century Massachusetts." *Jour-
       nal of Economic History* 36, 1 (March 1976): 263-266.

1831. Fleming, C.G. "The Plight of Black Educators in Post-
       war Tennessee, 1865-1920." *Journal of Negro History*
       64, 4 (Fall 1976): 355-364.

1832. Franklin, B. "Proposals Relating to the Education of
       Youth in Pennsylvania." *Journal of General Educa-
       tion* 28, 3 (Fall 1976): 256-261.

1833. Gordon, M.M. "Patriots and Christians: A Reassessment
       of Nineteenth-Century School Reformers (Massachusetts
       1830-1837)." *Journal of Social History* 11, 4 (Summer
       1978): 554-573.

1834. ————. *Union with the Virtuous Past: Development of
       School Reform in Massachusetts, 1789-1837.* Pitts-
       burgh, PA: University of Pittsburgh Press, 1974.

1835. Grossman, L. "George T. Downing and the Desegregation
       of Rhode Island Public Schools, 1855-1866." *Rhode
       Island History* 36, 4 (November 1977): 99-105.

1836. Holland, A.F., and G.R. Kremer. "Some Aspects of Black
       Education in Reconstruction Missouri: An Address by
       Richard B. Foster." *Missouri Historical Review*
       70, 2 (January 1976): 184-198.

1837. Johnson, K. "Elementary and Secondary Education in
       Iowa, 1890-1900: A Time of Awakening, Part I."
       Annals of Iowa 45, 7 (Fall 1979): 87-109.

1838. ————. "Elementary and Secondary Education in Iowa,
       1890-1900: A Time of Awakening, Part II." *Annals
       of Iowa* 45, 7 (Winter 1980): 171-195.

1839. Kousser, J.M. "Progressivism--For Middle Class Whites
       Only: North Carolina Education 1880-1910." *Journal
       of Southern History* 46, 2 (May 1980): 164-194.

1840. Lauderdale, W.B. "A Progressive Era for Education in
       Alabama, 1935-1951." *Alabama Historical Quarterly*
       37, 1 (Spring 1975): 38-63.

1841. Lee, A.E. "The Decline of Radicalism and Its Effect
       on Public Education in Missouri." *Missouri Historical
       Review* 74, 1 (October 1979): 1-20.

1842. Maxcy, S.J. "Idea of Consolidation in Southern Education during the Early Decades of the Twentieth Century." *Peabody Journal of Education* 53, 3 (April 1976): 216-222.

1843. Messner, W.F. "Black Education in Louisiana 1863-1865." *Civil War History* 22, 1 (March 1976): 41-59.

1844. Raichle, D.R. "The Abolition of Corporal Punishment in New Jersey Schools." *History of Childhood Quarterly* 2, 1 (Summer 1974): 53-78.

1845. Smith, S.D. *Schools and Schoolmen: Chapters in Texas Education, 1870-1900.* Denton, TX: North Texas State University Press, 1974.

1846. Taggart, R.J. "Philanthropy and Black Public Education in Delaware, 1918-1930." *Pennsylvania Magazine of History and Biography* 103, 4 (October 1979): 467-483.

1847. Tomberlin, J.A. "Florida and the School Desegregation Issue, 1954-1959: A Summary View." *Journal of Negro Education* 43, 4 (Fall 1974): 457-467.

1848. White, K.B. "The Alabama Freedman's Bureau and Black Education: The Myth of Opportunity." *Alabama Review* 34, 2 (April 1981): 107-124.

1849. Wilson, K. "Education as a Vehicle of Racial Control: Major General N.P. Banks in Louisiana, 1863-1864." *Journal of Negro Education* 50, 2 (Spring 1981): 156-170.

1850. Wollenberg, C. "Mendez vs. Westminister: Race, Nationality and Segregation in California Schools." *California Historical Quarterly* 53, 4 (Winter 1974): 317-332.

City Public Schools

The East

1851. Andrews, A.R. "The Baltimore School Building Program, 1870 to 1900: A Study of Urban Reform." *Maryland History Magazine* 70, 3 (Fall 1970): 260-274.

1852. Berrol, S.C. "Education and Economic Mobility: The Jewish Experience in New York City, 1880-1920." *American Jewish Historical Quarterly* 65, 3 (March 1976): 257-271.

1853. Bloom, L.A. "A Successful Jewish Boycott of the New York City Public Schools--Christmas 1906." *American Jewish History* 70, 2 (December 1980): 180-188.

1854. Brenzel, B. "Lancaster Industrial School for Girls: A Social Portrait of a Nineteenth-Century Reform School for Girls." *Feminist Studies* 3, 3 (Fall 1975): 40-53.

1855. Carroll, R. "America's First Public High School; English High School, Boston." *American Education* 11, 7 (August 1975): 20-21.

1856. Cleveland, G.O. "An Educational Profile of Erie, Pennsylvania, 1795-1850." *Journal of Erie Studies* 7 (Fall 1978): 35-53.

1857. Courchesne, G.L. "The Growth of Public Education in Holyoke, 1850 to 1873." *Historical Journal of Western Massachusetts* 7 (June 1979): 15-24.

1858. Cuban, L. "Reform by Fiat: The Clark Plan in Washington, 1970-1972." *Urban Education* 9, 1 (April 1974): 8-34.

1859. Fishbone, R.B. "The Shallow Boast of Cheapness: Public School Teaching as a Profession in Philadelphia, 1865-1890." *Pennsylvania Magazine of History and Biography* 103, 1 (January 1979): 66-84.

1860. Foley, F.J. "The Failure of Reform: Community Control and the Philadelphia Public Schools." *Urban Education* 10, 4 (January 1976): 389-414.

1861. Hammach, D.C. *Participation in Major Decisions in New York City, 1890-1900: The Creation of Greater New York and the Centralization of the Public School System.* New York: Columbia University Press, 1973.

1862. Hornburger, J.M. "Deep Are the Roots: Busing in Boston." *Journal of Negro Education* 45, 3 (Summer 1976): 235-245.

1863. Howard, R.W. *Education and Ethnicity in Colonial New York, 1664-1763: A Study in the Transmission of Culture in Early America.* Unpublished dissertation. University of Tennessee, 1978.

1864. Johnson, M. "Antoinette Brevost: A School-Mistress in Early Pittsburgh." *Winterthur Portfolio* 15 (Summer 1980): 151-168.

1865. Krasner, M. "The Politics of Public Education in New York City, 1970-1975." *Urban Education* 14, 4 (January 1980): 388-414.

1866. Entry deleted.

1867. Leon, W. *High School: A Study of Youth and Community
       in Quincy, Massachusetts.* Unpublished dissertation.
       Howard University, 1979.

1868. Levesque, G.A. "Before Integration: The Forgotten
       Years of Jim Crow Education in Boston." *Journal of
       Negro Education* 48, 2 (Spring 1979): 113-125.

1869. ————. "White Bureaucracy, Black Community: The Con-
       test over Local Control of Education in Antebellum
       Boston." *Journal of Educational Thought* 11, 2
       (August 1977): 140-155.

1870. Mabee, C. "Long Island's Black School War and the De-
       cline of Segregation in New York State." *New York
       History* 58, 4 (October 1977): 385-411.

1871. Marks, B.E. "Liberal Education in the Gilded Age:
       Baltimore and the Creation of the Manual Training
       School." *Maryland History Magazine* 74, 3 (Fall
       1979): 238-252.

1872. Meiring, B.J. *Educational Aspects of the Legislation
       of the Councils of Baltimore 1829-1884.* New York:
       Arno Press, 1978. Reprint.

1873. Modell, J. "An Ecology of Family Divisions: Suburbani-
       zation, Schooling, and Fertility in Philadelphia,
       1880-1920." *Journal of Urban History* 6, 4 (August
       1980): 397-417.

1874. Murphy, J.N. *Schools and Schooling in Eighteenth-
       Century Philadelphia.* Bryn Mawr, PA: Bryn Mawr
       College Press, 1977.

1875. O'Conner, C.A. "Setting a Standard for Suburbia:
       Innovation in the Scarsdale Schools, 1920-1930."
       History of Education Quarterly 20, 3 (Fall 1980):
       295-311.

1876. Pablo, J.M. *Washington, D.C. and Its School System,
       1900-1906.* Washington, D.C.: Georgetown University
       Press, 1975.

1877. Price, E.J. "School Segregation in Nineteenth-Century
       Pennsylvania." *Pennsylvania History* 43, 2 (April
       1976): 121-137.

1878. Proctor, R. *Racial Discrimination against Black
       Teachers and Black Professionals in the Pittsburgh
       School System, 1834-1973.* Unpublished dissertation.
       University of Pittsburgh, 1979.

1879. Ramsey, J.G. "The Education of Black Philadelphia:
       The Social and Educational History of a Minority
       Community 1900-1950." *Urban Education* 15, 2 (July
       1980): 251-254.

1880. Schiff, M. "Community Control of Inner City Schools
       and Educational Achievement (New York City)."
       Urban Education 10, 4 (January 1976): 415-428.

1881. Schumacher, C.S. *School Attendance in Nineteenth-
       Century Pittsburgh: Wealth, Ethnicity, and Occupa-
       tional Mobility of School Age Children 1855-1865.*
       Pittsburgh, PA: University of Pittsburgh Press,
       1977.

1882. Segal, B. "Jewish Schools and Teachers in Metropoli-
       tan Providence the First Century." *Rhode Island
       Jewish Historical Notes* 7 (November 1977): 410-419.

1883. Seller, M. "The Education of Immigrant Children in
       Buffalo, New York 1890-1916." *New York History*
       52, 2 (April 1976): 183-199.

1884. Shade, W.G. "The Working Class and Educational Reform
       in Early America: The Case of Providence, Rhode
       Island." *Historian* 39, 1 (November 1976): 1-23.

1885. Stack, J.F. "Ethnicity, Racism and Busing in Boston:
       The Boston Irish and School Desegregation." *Ethnicity*
       6, 1 (March 1979): 21-28.

1886. Thomas, B.C. "Public Education and Black Protest in
       Baltimore, 1865-1900." *Maryland Historical Magazine*
       71, 3 (Fall 1976): 381-391.

1887. Ueda, R. "Suburban Social Change and Educational
       Reform: The Case of Somerville, Massachusetts."
       Social Science History 3, 3/4 (1979): 167-203.

1888. Wood, D.M. "A Case Study in Local Control of Schools:
       Pittsburgh, 1900-1906." *Urban Education* 10, 1
       (April 1975): 7-26.

The Midwest

1889. Angus, D.I. "Detroit's Great School Wars: Religion
       and Politics in a Frontier City, 1842-1853." *Michi-
       gan Academician* 12, 3 (Winter 1980): 261-280.

1890. Bosco, J.J., and S.S. Robin. "White Flight from
       Busing? A Second, Longer Look (Kalamazoo, Michi-
       gan)." *Urban Education* 11, 3 (October 1976): 263-
       274.

1891. Calkins, D.I. "Black Education in Nineteenth-Century
       Cincinnati." *Cincinnati Historical Society Bulletin*
       38, 2 (Summer 1980): 115-128.

1892. Candelero, D. "The Chicago School Board Crises of
       1907." *Journal of the Illinois State Historical
       Society* 68, 3 (November 1975): 396-406.

1893. Cohen, R., and R. Mohl. *The Paradox of Progressive
       Education: The Gary Plan and Urban Schooling 1900-
       1940.* Port Washington, NY: Kennikat Press, 1979.

1894. Cortinovis, I.E. "Documenting an Event with Manu-
       scripts and Oral History: The St. Louis Teachers'
       Strike, 1973." *Oral History Review* 3 (1974): 59-63.

1895. Daniel, P.T.K. "A History of Discrimination Against
       Black Students in Chicago Secondary Schools." *His-
       tory of Education Quarterly* 20, 2 (Summer 1980):
       147-162.

1896. Dawson, L.R. "H.G. Wells of Kalamazoo: Pioneer Advo-
       cate of Education." *Old Northwest* 6, 1 (Spring 1980):
       43-61.

1897. Dye, C.M. "Calvin Woodward, Manual Training and the
       Saint Louis Public Schools." *Bulletin of the Missouri
       Historical Society* 31, 2 (January 1975): 111-135.

1898. Homel, M.W. "The Politics of Public Education in
       Black Chicago, 1910-1941." *Journal of Negro Educa-
       tion* 45, 2 (Spring 1976): 179-191.

1899. Johnson, R.M. "Politics and Pedagogy: The 1892 Cleve-
       land School Reform." *Ohio History* 84, 4 (Autumn
       1975): 196-206.

1900. Kirby, J.R. "Fort Wayne Common School Crusaders: The
       Construction of the Clay Street School 1854-1857."
       Old Fort News 44, 2 (1980): 50-60.

1901. ————. "Fort Wayne Common School Crusaders: The
       First Year for Free Schooling, April 1853-March
       1854." *Old Fort News* 42, 1 (1979): 13-25.

1902. Miggins, E.M. *The Cleveland Foundations and Cleveland
       Public School Reform during the Progressive Period.*
       Cleveland, OH: Case Western Reserve University Press,
       1975.

1903. Miller, J. "Public Elementary Schools in Cincinnati,
       1870-1914." *Cincinnati Historical Society Bulletin*
       38, 2 (Summer 1980): 83-95.

1904. ———. "Urban Education and the New City: Cincinnati's Elementary Schools, 1870-1914." *Ohio History* 88, 2 (Spring 1979): 152-172.

1905. Pennoyer, J.C. *The Harper Report of 1899: Administrative Progressivism and the Chicago Public Schools.* Unpublished dissertation. University of Denver, 1979.

1906. Peterson, P.E. *School Politics: Chicago Style.* Chicago, IL: University of Chicago Press, 1977.

1907. Reese, W.J. "Progressive School Return in Toledo: 1898-1921." *Northwest Ohio Quarterly* 47, 2 (Spring 1975): 44-59.

1908. Smith, J.K. "Progressive School Administration: Ella Flagg Young and the Chicago Schools, 1905-1915." *Journal of the Illinois State Historical Society* 73, 1 (Spring 1980): 27-44.

1909. Trebling, H.M. "The Chicago School versus Public Utility Regulation." *Journal of Economic Issues* 10, 2 (March 1976): 97-126.

The South

1910. Lord, J.D., and J.C. Catau. "School Desegregation, Busing, and Suburban Migration (Charlotte, N.C.)." *Urban Education* 11, 3 (October 1976): 275-294.

1911. Muller, M.L. "New Orleans Public School Desegregation." *Louisiana History* 17, 1 (Winter 1976): 69-88.

1912. Newman, J.W. "The Social Origins of Atlanta's Teachers, 1881, 1896, 1922." *Urban Education* 11, 1 (April 1976): 115-122.

1913. Racine, P.N. "A Progressive Fights Efficiency: The Survival of Willis Sutton, School Superintendent (Atlanta)." *South Atlantic Quarterly* 76 (Winter 1977): 103-116.

1914. ———. "Willis Anderson Sutton and Progressive Education, 1921-1943." *Atlanta Historical Bulletin* 20 (Spring 1976): 9-23.

1915. Riter, M.I.C. "Teaching School at Old Fort Laramie." *Annals of Wyoming* 51, 2 (Fall 1979): 24-25.

1916. Van Meeter, S. "Black Resistance to Segregation in the Wichita Public Schools, 1870-1912." *Midwest Quarterly* 20, 1 (Autumn 1978): 64-77.

1917. Wallace, D. "Orval Faubus: The Central Figure at Little

Rock Central High School." *Arkansas Historical Quarterly* 39, 2 (Summer 1980): 314-329.

1918.   White, A.O.   "Black Boycott: Gainesville, Florida."
        Urban Education 9, 4 (January 1975): 309-324.

1919.   Wright, C.T.   "The Development of Public Schools for
        Blacks in Atlanta, 1872-1900." *Atlanta Historical
        Bulletin* 21 (Spring 1977): 115-128.

1920.   Yancy, D.C.   "William Edward Burghardt Dubois' Atlanta
        Years." *Journal of Negro History* 63, 1 (January
        1978): 59-67.

The West

1921.   Biebel, C.D.   "Cultural Change on the Southwest Fron-
        tier: Albuquerque Schooling 1870-1895." *New Mexico
        Historical Review* 55, 3 (July 1980): 209-230.

1922.   Boulton, S.W.   "Desegregation of the Oklahoma City School
        System." *Chronicle of Oklahoma* 58, 2 (Summer 1980):
        192-220.

1923.   Carleton, D.E.   "McCarthyism in Local Elections: The
        Houston School Board Election of 1952." *Houston Re-
        view* 3 (Winter 1981): 168-177.

1924.   Gonzales, G.G.   "Educational Reform in Los Angeles and
        Its Effect upon the Mexican Community, 1900-1930."
        Exploring Ethnic Studies 1 (July 1978): 5-26.

1925.   ————.   "Racism, Education and the Mexican Community
        in Los Angeles, 1920-1930." *Societas* 4, 4 (Autumn
        1974): 287-301.

1926.   Kirp, D.L.   "Race, Politics, and the Courts: School De-
        segregation in San Francisco." *Harvard Education Re-
        view* 46 (November 1976): 572-611.

1927.   Pieroth, D.   "With All Deliberate Caution: School In-
        tegration in Seattle, 1954-1968." *Pacific Northwest
        Quarterly* 73, 2 (April 1982): 50-61.

1928.   Shrader, V.L.   "Ethnicity, Religion, and Class: Pro-
        gressive School Reform in San Francisco." *History of
        Education Quarterly* 20, 4 (Winter 1980): 385-404.

1929.   Strober, M.H., and L. Best.   "The Female/Male Salary
        Differential in Public Schools: Some Lessons from San
        Francisco, 1879." *Economic Inquiry* 17, 2 (April
        1979): 218-236.

1930.   Weiss, M.   "Education, Literacy, and the Community of
        Los Angeles in 1850." *Southern California Quarterly*
        60 (Summer 1978): 117-142.

MOVING GOODS AND PEOPLE

Urban sprawl has aggravated the congested transportation systems of most cities. Moving goods, services, and people to and from various locations in the urban maze is at best cumbersome and frequently only a tribute to good luck. Mass transit, a feature common to cities before the widespread use of the automobile, is now a struggling infant in comparison to the dependency upon that individual motorized vehicle. This decline of municipal transit systems and the subsequent dominance of the automobile is detailed in Mark Foster's *From Streetcar to Superhighway: American City Planners and Urban Transportation 1900-1940* (1981). Foster contends that a perception was created that the automobile was more flexible and economical than the then-in-use mass transit systems. This perception hastened the change in urban transportation mode from mass transit to one which offered a private and convenient travel possibility more appealing to Americans. An earlier work, *American Ground Transport* (1974) by Bradford C. Snell, had outlined a different theory for the rise of automobile usage. Snell believed that a conspiracy of automobile, highway, and related interest groups lobbied in local and national government agencies to dismantle fixed-rail mass transit. This work was supported in *The Political Economy of Urban Transportation* (1977) by Delbert A. Taebel and James V. Cornehls.

The use of the automobile and superhighway for the transportation of goods and people has been the subject of several recent studies. *Interstate Express Highway Politics 1941-1956* (1979) by Mark H. Rose is an excellent study of the creation of the American expressway systems. Specific focus upon the problems of highway creation and maintenance in the urban environment is found in *L.A. Freeway* (1981) by David Brodsley. Both these studies serve as good introductions to the advantages and disadvantages of the automobile as the principal means of urban transport.

Mass transit, however, is the present heralded cure for the transportation problems which confront large metropolitan areas. For several decades there has been a reliance upon both the automobile and mass transit systems to move significant

but different portions of the urban populace. Solutions--
bus, subway, monorail, and even people-mover--have been tried,
adopted, and then adapted into coexistence with the automobile.
Cities have become crowded with various types of transportation
which need planning, directing, supervising, and funding.
Municipal governments have become caught in a wheel of transportation alternatives which seem to provide only temporary
relief from transportation problems but claim an ever-in-
creasing amount of time and money to operate. A recent study,
*Moving the Masses: Urban Public Transit in New York, Boston,
and Philadelphia, 1880-1912* (1981) by Charles W. Cheane examines the previous unique mass transit systems which were
operated in large eastern metropolitan areas before the advent
of the automobile. Though the time period is distant, this
study is useful for its presentation of problems confronting
city governments in transportation and the manner in which
solutions, successful and otherwise, were implemented.

Urban transportation is a subject area relating to the
social, historical, and physical sciences. Research has been
conducted in the field by historians, sociologists, economists, civil engineers, and government managers. The following list of books and journal articles is representative of
the research conducted during the last ten years.

1931. Albert, J., and S. Banton. "Urban Spatial Adjustments
      Resulting from Rising Energy Costs." *Annals of Re-
      gional Science* 12, 2 (July 1978): 64-71.

1932. Allen, G.R., et al. "Status of Intercity Bus Service
      in Virginia and Anticipated Impacts of Regulatory
      Reform." *Transportation Quarterly* 36, 4 (October
      1982): 597-615.

1933. Alpert, M., and L. Golden. "The Marketing of Mass
      Transportation to Diverse Groups within a Community."
      Journal of Urban Analysis 5, 2 (1978): 285-312.

1934. Altshuler, A. *The Urban Transportation System: Poli-
      tics and Policy Innovation.* Cambridge, MA: MIT
      Press, 1979.

1935. Appleyard, D. "Livable Streets: Protected Neighbor-
      hoods." *American Academy of Political and Social
      Science Annals* 451 (September 1980): 106-107.

1936. Baehr, G.T. "Conrail Bows Out of Urban Transit."
      Mass Transit 10 (January 1983): 50-53.

1937. Baldassare, M., et al. "Urban Service and Environmental
      Stressor: The Impact of the Bay Area Rapid Transit
      System." *Environment and Behavior* 11, 4 (1979): 435-
      450.

1938. Balshone, B. *Bicycle Transit; Its Planning and Design.* New York: Praeger, 1975.

1939. Barker, W. "A Simulation of Commuter Rail Possibilities." *Computers, Environment and Urban Systems* 3, 2/3 (1978): 123-129.

1940. Bartlett, P. "Public Policy and Private Choice: Mass Transit and the Automobile in Chicago between the Wars." *Business History Review* 44, 4 (Winter 1975): 473-497.

1941. Bauer, A.E. "Solving Transportation Problems in the Federal System: Is There a Role for State and Local Governments." *Publius* 8, 2 (1978): 59-76.

1942. Berkman, H. "Some Perspectives on Transportation in the Next Decade." *Traffic Quarterly* 34, 1 (January 1980): 143-154.

1943. Bland, W.R. "Urban Transportation Planning in California: An Evaluation." *Professional Geographer* 30, 2 (1978): 156-161.

1944. Bollen, K. "Suicidal Motor Vehicle Fatalities in Detroit: A Replication." *American Journal of Sociology* 87, 2 (September 1981): 404-412.

1945. Braun, R. "Bring Back the Trolley Bus? Would the Trolley Bus Improve Downtown Circulation and Environment?" *Metro* 18 (January/February 1977): 15-18.

1946. Brodsley, D. *L.A. Freeway: An Appreciative Essay.* Berkeley, CA: University of California Press, 1981.

1947. "Bus Drivers Key to Fuel Efficiency." *Mass Transit* 10 (January 1983): 12-13.

1948. "Capacity Characteristics of Downtown Bus Streets." *Transportation Quarterly* 36, 4 (October 1982): 617-630.

1949. Castle, G., and L. Slade. "New Rapid Rail Systems: Will They Do the Job?" *Planning* 43, 7 (July 1977): 16-19.

1950. Chall, D. "Chance AMTV (Articulated Modular Transit Vehicle) Ready for Mass Transit Introduction." *Metro* 22 (January/February 1981): 10-12.

1951. Chernoff, M. "The Effects of Superhighways in Urban Areas." *Urban Affairs Quarterly* 16, 3 (March 1981): 317-336.

1952. Chicago Area Transportation Study. *Year 2000 Planning*

Process. Chicago, IL: Chicago Area Transportation Study, 1982.

1953. Cohn, L.F. "Environmental Action Planning for Transportation Project Development." *Transportation Quarterly* 36, 4 (October 1982): 503-525.

1954. Cooke, H. "Urban Passenger Transportation Systems: Some Selection Criteria." *Urban Forum/Colloque Urbain* 4 (January-February-March 1979): 4-9.

1955. Cottrell, B.H., and M.J. Demetsky. "Preliminary Design and Evaluation of Coordinated Public Transportation Services." *Traffic Quarterly* 35, 1 (January 1981): 143-162.

1956. Daniels, P.W., and A.M. Warnes. *Movement in Cities: Spatial Perspectives on Urban Transport and Travel.* New York: Methuen, 1980.

1957. Davies, R. *The Age of Asphalt: The Automobile, the Freeway, and the Condition of Metropolitan America.* Philadelphia: Lippincott, 1975.

1958. Debski, M. "A Demand-Response Bus System: The Valley Transit District." *Traffic Quarterly* 30, 3 (July 1976): 431-447.

1959. Deen, T. "The Potential for Transit Standards." *Traffic Quarterly* 31, 1 (January 1977): 119-137.

1960. Domoro, H. "The San Diegans: A Cinderella Story." *Mass Transit* 6 (November 1979): 18-22.

1961. Edwards, J.L. "Future Relationships between Long and Short Urban Transportation Planning." *Traffic Quarterly* 32, 4 (October 1978): 531-544.

1962. Feaver, D. "Metro: The Growing Pains of Success." *Mass Transit* 6 (July 1979): 16-21.

1963. Fisch, O. "The Social Cost of Through Traffic--Contribution to the Suburban-Central City Exploitation Thesis." *Regional Science and Urban Economics* 5, 2 (May 1975): 263-277.

1964. Fischler, S. *Uptown, Downtown: A Trip through Time on New York's Subways.* New York: Hawthorne Books, 1976.

1965. Foster, J., and M. Schmidt. "Rail Terminals in the Urban Environment." *Transportation Journal* 14, 1 (Fall 1975): 21-28.

1966. Foster, M. "City Planners and Urban Transportation: The American Response, 1900-1940." *Journal of Urban History* 5, 3 (1979): 365-396.

1967. ————. *From Streetcar to Super Highway: American City Planners and Urban Transportation*. Philadelphia, PA: Temple University Press, 1981.

1968. Fravel, F., et al. "Free-Wheeling Los Angeles Begins Third Century in Stop and Go Traffic." *Mass Transit* 8 (September 1981): 6-11.

1969. Gray, G. *Public Transportation: Planning, Operations, and Management*. Englewood Cliffs, NJ: Prentice-Hall, Inc., 1979.

1970. Gray, P., and O. Helmer. "The Use of Futures Analysis for Transportation Research Planning." *Transportation Journal* 15, 2 (Winter 1976): 5-12.

1971. Guest, A., and C. Cluett. "Workplace and Residential Location: A Pushpull Model." *Journal of Regional Science* 16, 3 (1976): 399-410.

1972. Haines, R. "The Politics of Economics in Transit Planning." *Urban Affairs Quarterly* 14, 1 (September 1978): 55-77.

1973. Hall, E. "Urban Transportation Financing: The Phoenix Program." *Traffic Quarterly* 31, 2 (April 1977): 275-286.

1974. Hamer, A. *The Selling of Rail Rapid Transit: A Critical Look at Urban Transportation Planning*. Lexington, MA: Lexington Books, 1976.

1975. Hamilton, N. *Governance of Public Enterprise: A Case Study of Urban Mass Transit*. Lexington, MA: Lexington Books, 1981.

1976. Harbeson, R. "Social Welfare and Economic Efficiency in Transport Policy." *Land Economics* 7, 1 (February 1977): 97-105.

1977. Harmatuck, D. "Referendum Voting as a Guide to Public Support for Mass Transit." *Traffic Quarterly* 30, 3 (July 1976): 347-369.

1978. Harmen, C., and J. Cudlin. "Pneumatic Tube Transportation." *High Speed Ground Transportation Journal* 9, 1 (Fall 1975): 181-189.

1979. Hassell, J. "How Effective Has Urban Transportation Planning Been?" *Traffic Quarterly* 34, 1 (January 1980): 5-20.

1980. Hawley, Miriam. "BART and the San Francisco Bay Area." *Planning and Administration* 2 (Autumn 1975): 25-32.

1981. Hayes, D. "Rapid Transit Financing: Use of the Special
       Assessment." *Stanford Law Review* 29, 4 (April 1977):
       795-818.

1982. Hazard, J. "National Transportation Policy Adminis-
       tration: Transitional Lessons from Home and Abroad."
       Transportation Journal 16, 4 (Summer 1977): 5-19.

1983. Heggie, I. "Consumer Response to Public Transport
       Improvements and Car Restraint: Some Practical
       Findings." *Policy and Politics* 5 (June 1977): 47-64.

1984. Hensher, D. "Urban Transportation Planning--The
       Changing Emphasis." *Socio-Economic Planning Sciences*
       13, 2 (1979): 95-104.

1985. Herbert, B. "Urban Morphology and Transportation."
       Traffic Quarterly 30, 4 (October 1976): 633-649.

1986. Hebert, R. "Commuting: Busway Big Success in LA."
       Mass Transit 4 (February 1977): 22-31.

1987. ———. "Minibuses a Downtown Attraction." *Mass
       Transit* 4 (April 1977): 30-31.

1988. Herbert, S.L., and R.A. Weant. *Urban Transportation:
       Perspectives and Prospects*. Westport, CT: Eno Founda-
       tion for Transportation, 1982.

1989. Hicks, R., and M. Kobran. "Transportation's Role in
       Downtown Detroit Revitalization." *Traffic Quarterly*
       33, 3 (July 1979): 331-346.

1990. Higgins, T.J. "Coordinating Buses and Rapid Rail in
       the San Francisco Bay Area: The Case of Bay Area
       Rapid Transit." *Transportation* 10, 4 (1981): 357-
       371.

1991. Huth, T., and R. Kaul. "Chicago's El Rattles On--and
       Outlasts Its Critics." *Historic Preservation* 32
       (January/February 1980): 2-10.

1992. Jackson, R. "The Cost and Quality of Paratransit
       Service for the Elderly and Handicapped." *Transpor-
       tation Quarterly* 36, 4 (October 1982): 527-540.

1993. Jennrich, J. "Passenger Transportation: People in Per-
       petual Motion." *Nation's Business* 67, 5 (May 1979):
       85-91.

1994. ———. "Transportation 2000: How America Will Move
       Its People and Products." *Nation's Business* 67, 11
       (November 1979): 34-40.

1995. Jones, I. *Urban Transport Appraisal*. New York: John
       Wiley and Sons, 1977.

1996. Kim, T., and W. Volk. "Creating an Upward Cycle in Urban Transit Ridership: A Case Study." *Traffic Quarterly* 33, 4 (October 1979): 501-510.

1997. Kleine, D. "Factors Contributing to the Success of Community-Chartered Commuter Bus Service." *Urban Land* 34, 11 (November 1975): 16-19.

1998. Kuner, R. "Alternative Analysis for Arterial Streets." *Traffic Quarterly* 33, 3 (July 1979): 459-472.

1999. Lalani, N. "Three Ways to Use Pavement Stripes to Cut Accident Rates." *American City and County* 97, 9 (September 1982): 41.

2000. Lane, B. "Conference Takes Second Look at Trolley Buses." *Mass Transit* 9 (December 1982): 12-13.

2001. Langdon, F. "Monetary Evaluation of Nuisance from Road-Traffic Noise: An Exploratory Study." *Environment and Planning* 10, 8 (1978): 1015-1034.

2002. Lea, N., and L. Suen. "The Taxi: An Urban Transportation Resource." *Urban Forum/Colloque Urbain* 3 (March/April 1978): 16.

2003. Levinson, H.S. "Travel Restraints in City Centers: The American Experience." *Transportation Quarterly* 37, 2 (1983): 277-288.

2004. Lopata, H.Z. "The Chicago Woman: A Study of Patterns of Mobility and Transportation." *Signs* 5 (Spring 1980): 161-169.

2005. Lovely, M. "Public Transit and Downtown Development." *Urban Land* 38, 11 (November 1979): 14-22.

2006. Lowry, G. *Streetcar Man: Tom Lowry and the Twin City Rapid Transit Company.* Minneapolis, MN: Lorner Publications, 1979.

2007. Mamon, J., and H. Marshall. "The Use of Public Transportation in Urban Areas: Toward a Causal Model." *Demography* 14, 1 (February 1977): 19-31.

2008. Mayer, H. "Milwaukee's Accessibility Program—A Lot of Work for a Little Mobility." *Metro* (March/April 1980): 35.

2009. McCulley, B. "Anatomy of a Bus Shelter." *American City and County* 96, 2 (February 1981): 66-68.

2010. McDermott, D. "An Alternate Framework for Urban Goods Distribution: Consolidation." *Transportation Journal* 14, 1 (Fall 1975): 29-39.

2011. ————. "Urban Goods Movement: State of the Art and
       Future Possibilities." *Transportation Journal* 19, 2
       (Winter 1980): 34-40.

2012. McLear, P.E. "The Galena and Chicago Union Railroad:
       A Symbol of Chicago's Economic Maturity." *Journal
       of the Illinois State Historical Society* 73, 1
       (Spring 1980): 17-26.

2013. Melton, L. "Transport Regulation: An Effective Tool
       of Public Policy." *Transportation Journal* 17, 3
       (Spring 1978): 86-94.

2014. Meyer, J. "National Transportation Policy Plans--A
       Critique." *Transportation Journal* 15, 2 (Winter
       1976): 30-34.

2015. Meyer, M.D., and P. Belobaba. "Contingency Planning
       for Response to Urban Transportation System Disrup-
       tion." *American Planning Association Journal* 48, 4
       (Autumn 1982): 454-456.

2016. Miller, E. "Effects of City Size and Population Den-
       sity on Highway Wage and Needs." *American Journal
       of Economics and Sociology* 37 (July 1978): 295-308.

2017. Miller, G., and M. Green. "Commuter Van Programs--An
       Assessment." *Traffic Quarterly* 31, 1 (January 1977):
       33-57.

2018. Miller, J. "An Evaluation of Allocation Methodologies
       for Public Transportation Operating Assistance."
       Transportation Journal 18, 1 (Fall 1979): 40-49.

2019. Morris, B. "Marketing." *Mass Transit* 5 (February
       1978): 6-11.

2020. Morris, R. "Urban Sounding Board--Do Freeways Help
       Downtown?" *Nation's Cities* 13, 11 (November 1975):
       38-41.

2021. Nairn, J. "Mass Transit: A Look at Trendsetting De-
       signs That Keep Us Moving." *Architectural Record*
       137 (July 1979): 115-130.

2022. Nelson, H., and W. Barnes. "Chicago Urban Transporta-
       tion Planning." ASCE. *Journal of the Urban Planning
       and Development Division* 103 (July 1977): 53-67.

2023. Nelson, K., and W. Nevel. "Cost-Effectiveness Analysis
       of Public Transit Systems." *Traffic Quarterly* 33, 2
       (April 1979): 241-252.

2024. Niedowski, R., and D. Low. "Survey Procedures for the

Boston Central Artery Origin-Destination Study."
Traffic Quarterly 33, 4 (October 1979): 555-576.

2025. Norris, G., and N. Nihan. "Subarea Transportation
Planning: A Case Study." *Traffic Quarterly* 33, 4
(October 1979): 589-605.

2026. Nupp, B. "Trends and Choices for Intercity Passenger
Transportation in a Era of Research Stringency."
Transportation Journal 19, 4 (Summer 1980): 48-52.

2027. Orski, C.K. "Private Enterprise and Public Transportation." *Vital Speeches* 49, 1 (October 1982):
18-22.

2028. Owen, W. "An Urban Transformation through Transportation." *Nation's Cities* 15, 5 (May 1977): 15-20.

2029. "People vs. Cars--An Urban Dilemma." *American City
and County* 91, 8 (August 1976): 33-35.

2030. Petrocelli, J., and T. Bell. "Assessing Demand for
Ride-Sharing Services." *Traffic Quarterly* 31, 1
(January 1977): 59-76.

2031. Piper, R. "Transit Strategies for Suburban Communities." *American Institute of Planners Journal* 43
(October 1977): 380-385.

2032. Polonsky, S. "Serving Transportation Needs of the
Elderly: An Overview." *Traffic Quarterly* 32, 4
(October 1978): 621-633.

2033. Portararo, A. "A Mass Transit Model Urban Transportation." *Traffic Quarterly* 30, 4 (October 1976):
561-576.

2034. Poulton, M.C. "The Best Pattern of Residential Streets."
American Planning Association Journal 48, 4 (Autumn
1982): 466-480.

2035. "Public Transit and Downtown Development: Mobility
and Growth Must Go Together." *Metro* 21 (May/June
1980): 48-54.

2036. Pucker, J. "Effects of Subsidies on Transit Costs."
Transportation Quarterly 36, 4 (October 1982): 549-
562.

2037. ———. "Who Benefits from Transit Subsidies? Recent
Evidence from Six Metropolitan Areas." *Transportation
Research* 17A, 1 (1983): 39-50.

2038. Pushkarev, B. *Public Transportation and Land Use Policy.*
Bloomington, IN: Indiana University Press, 1977.

2039. Rallis, T. *Intercity Transport: Engineering and Planning*. New York: John Wiley and Sons, 1977.

2040. Rand, R., and W. Avery. "Applications of New Systems to Urban Transportation." *Traffic Quarterly* 31, 1 (January 1977): 97-117.

2041. Reed, A. "The Urban Mass Transportation Act and Local Labor Negotiation: The 13-C Experience." *Transportation Journal* 18, 3 (1979): 56-64.

2042. Richards, B. *Moving in Cities*. Boulder, CO: Westview Press, 1975.

2043. Ridings, R.L. "User Fee Pay for Clean Streets." *American City and County* 96, 2 (February 1981): 43-44.

2044. Ross, B. "Big Apple's Bus Service, Subways Are on the Slide." *Mass Transit* 8 (February 1981): 6-9+.

2045. Rowse, J. "Solving the Generalized Transportation Problem." *Regional Science and Urban Economics* 11, 1 (February 1981): 57-68.

2046. Sale, J. "Ride On!" *Practicing Planner* 8, 1 (March 1978): 24-29.

2047. Schoene, G.W., and G.W. Euler. "Let TRANSYT Speed Traffic Flow." *American City and County* 97, 12 (December 1982): 23-24.

2048. Schulz, D. "An Evolving Image of Long-Range Transportation Planning." *Traffic Quarterly* 33, 3 (July 1979): 443-457.

2049. Smith, A.N. "Blacks and the Los Angeles Municipal Transit System, 1945-1971." *Urbanism Past and Present* 6 (Winter/Spring 1981): 25-31.

2050. Smith, P.B. "Highway Planning in California's Motherlode: The Changing Townscape of Auburn and Nevada City." *California History* 59, 3 (1980): 204-221.

2051. Spielberg, F. *Transportation Improvement in Madison Wisconsin*. Washington, DC: Urban Land Institute, 1978.

2052. St. Clair, D.J. "The Motorization and Decline of Urban Public Transit, 1935-1950." *Journal of Economic History* 41, 3 (1981): 579-600.

2053. Stahurg, J.M. "Status Transition of Blacks and Whites in American Suburbs." *Sociological Quarterly* 23, 1 (Winter 1982): 79-94.

2054. Stanback, T.M., and T.J. Noyelle. *Cities in Transportation: Changing Job Structures in Atlanta, Denver, Buffalo, Phoenix, Columbus, Nashville, and Charlotte.* Totowa, NJ: Allanhad, Osmun and Company, 1982.

2055. Stano, M. "Consumption Externalities in Models of Urban Transportation." *Public Finance Quarterly* 5, 2 (April 1977): 231-246.

2056. Starkie, D. *Transportation Planning, Policy, and Analysis.* New York: Pergamon Press, 1976.

2057. Westcott, D. "Employment and Commuting Patterns: A Residential Analysis." *Monthly Labor Review* 102, 7 (July 1979): 3-9.

2058. Wheaton, W. "Residential Decentralization, Land Rents, and the Benefits of Urban Transportation Investment." *American Economic Review* 67, 1 (March 1977): 138-143.

2059. White, M. "A Model of Residential Location Choice and Commuting by Men and Women Workers." *Journal of Regional Science* 17, 2 (April 1977): 41-52.

2060. Weise, A. "... The Automobile Is the Enemy of the City." *Mass Transit* 4 (May 1977): 20-23.

2061. ———. "OMB: Urban Transit's Nagging Nemesis." *Mass Transit* 10 (January 1983): 16-19+.

2062. ———. "Reagan Administration: A Time of Not-So-Great Expectations for Transit." *Mass Transit* 8 (March 1981): 6-8+.

2063. Williams, E.W. "The National Transportation Policy Study Commission and Its Final Report: A Review." *Transportation Journal* 19, 3 (Spring 1980): 5-19.

2064. "Year Around DPM (Downtown People Mover)--A Reality with Winterization Technology." *Metro* 22 (January/February 1981): 30-36.

2065. Young, D. "Chicago: Stuck on the Loop." *Mass Transit* 7 (January 1980): 30-34.

2066. ———. "Chicago: The City of Continual Change." *Mass Transit* 5 (May 1978): 6-11.

2067. ———. "Davenport, Iowa: One Small System's Race Against Time." *Mass Transit* 9 (December 1982): 52-54+.

2068. ———. "Nortran Arises from Ashes of Chicago's Transit Crisis." *Mass Transit* 9 (October 1982): 26-27.

2069. ————. "St. Louis: Troubled Transit." *Mass Transit*
        6 (April 1979): 20-21.

2070. ————. "Seattle: Selling Transit with Innovations."
        Mass Transit 6 (December 1979): 16-24.

GEOGRAPHIC INDEX

The geographic index below contains listings for all fifty
states and major cities in each state. Additionally, there
are listings for cities which have received special treatment
in urban research and cities for which there has been a significant study conducted.

United States (General) 1-230, 877, 884-886, 888-890, 892-
    894, 898, 902, 908, 909, 912, 914, 916, 919, 922, 925-
    927, 934, 936, 937, 946, 948, 949, 951, 953, 960, 970,
    976, 982, 993, 995, 996, 1000, 1005, 1008, 1009, 1011,
    1012, 1014, 1015, 1019, 1021, 1026, 1027, 1031, 1033,
    1034, 1037, 1038, 1040, 1041, 1047, 1048, 1050, 1054-
    1060, 1063, 1066, 1067, 1072-1074, 1080, 1085, 1087,
    1089, 1090, 1094, 1095, 1104, 1106, 1111, 1114, 1123,
    1127-1129, 1131, 1134, 1139, 1142, 1158, 1164, 1166,
    1171-1174, 1176, 1179, 1186, 1187, 1194, 1206, 1209,
    1222, 1224-1230, 1241, 1249, 1252-1284, 1353-1372,
    1523-1613, 1678, 1770-1772, 1774-1780, 1782-1809, 1813-
    1822, 1931, 1933-1936, 1938, 1939, 1941, 1942, 1945,
    1947-1950, 1953-1959, 1961-1963, 1965-1967, 1969-1972,
    1974-1979, 1981-1985, 1987, 1988, 1992-2003, 2005,
    2007, 2009-2011, 2013-2021, 2023, 2025-2043, 2045-2048,
    2052-2064
East (General) 231-240, 1626, 1631, 1651
  Connecticut (General) 878, 1403, 1629, 1630
    Bridgeport 244, 269
    Hartford 272, 1124, 1386
    Milford 1418
    New Britain 1382
    New Haven 264, 1374, 1656
    Norwich 349, 1395
    Stamford 1635
  Delaware (General) 1846
    Cantwell's Bridge 386
    Wilmington 387
    Wooddale 315

East (cont'd)
  Maryland (General) 1175, 1161
    Annapolis 333
    Baltimore 281, 288, 292, 304, 313, 314, 322, 360, 370,
        397, 401, 903, 1075, 1201, 1394, 1414, 1422, 1423,
        1614-1616, 1628, 1639, 1646, 1660, 1851, 1871, 1872,
        1886
    Columbia 350, 1620
    Emmitsburg 381
    Hampden-Woodbury 287
  Massachusetts (General) 256, 258, 261, 271, 1219, 1286,
        1828-1830, 1833, 1834
    Boston 232, 238, 242-246, 250, 251, 253, 254, 257, 262,
        265, 266-268, 270, 273-275, 277, 278, 899, 924, 944,
        1069, 1375, 1388, 1389, 1412, 1417, 1430, 1439, 1650,
        1655, 1657, 1668, 1673, 1781, 1855, 1862, 1868, 1869,
        1885, 2024
    Haverhill 1285, 1287
    Holyoke 938, 1396, 1858
    Lowell 248, 252
    Lynn 247, 249, 259, 1288
    North Beverly 260
    Northrup 1424
    Quincy 1867
    Salem 255, 1377, 1392, 1419
    Somerville 227, 1887
    Worcester 1182
  New Hampshire
    Acworth 1427
    Concord 1409
  New Jersey (General) 236, 1051, 1844
    Atlantic City 344
    Chatham 407
    Jersey City 941, 1647, 1671
    Newark 559, 983, 1420
    Passaic 1289, 1290, 1634
    Patterson 1291
    Trenton 1177
  New York (General) 236, 343, 348, 383, 1251, 1425
    Albany 1662
    Beekmantown 408
    Buffalo 904, 1183, 1244, 1380, 1440, 1883
    Farmingdale 1189
    Jamestown 943, 1623
    New York City 231, 232, 238, 279, 282, 293, 294-297,
        301, 303, 310-312, 316, 319, 321, 323, 327, 330, 334,
        336, 341, 345-347, 353, 354, 358, 365-369, 373, 374,
        376-380, 389, 390, 392-395, 398-400, 402-405, 409-414,

880, 882, 887, 897, 900, 910, 918, 930, 959, 967, 968,
988, 998, 1007, 1022-1024, 1064, 1065, 1159, 1168,
1178, 1185, 1190, 1196, 1199, 1204, 1208, 1211, 1238,
1243, 1292, 1294, 1296-1300, 1378, 1381, 1385, 1387,
1393, 1397, 1398, 1400, 1402, 1405, 1408, 1410, 1413,
1421, 1433, 1437, 1619, 1621-1625, 1632, 1637, 1638,
1640-1645, 1648, 1649, 1653, 1658, 1659, 1663, 1664,
1670, 1672, 1674, 1676, 1781, 1852, 1853, 1861, 1863,
1870, 1880, 1964, 2044
Rochester 337, 1295, 1404, 1415
Scarsdale 1875
Schenectady 1293
Syracuse 364, 382, 384
Trumansburg 361
Utica 1426
Pennsylvania (General) 283, 317, 335, 339, 1301, 1832,
1877
Erie 1856
Lancaster 1854
Philadelphia 232, 283-285, 291, 298-300, 302, 305, 307,
309, 325, 328, 329, 352, 356, 357, 359, 363, 385, 391,
396, 896, 905, 942, 1013, 1032, 1071, 1108, 1152, 1153,
1203, 1302, 1379, 1383, 1390, 1391, 1406, 1428, 1429,
1433, 1434, 1438, 1617, 1618, 1652, 1654, 1665, 1666,
1859, 1860, 1873, 1874, 1879
Pittsburgh 231, 280, 324, 326, 331, 342, 362, 375, 388,
929, 950, 1092, 1160, 1163, 1191, 1213, 1416, 1431,
1435, 1436, 1667, 1669, 1675, 1864, 1878, 1881, 1888
Rhode Island (General) 1811, 1835, 1884
Newport 1384, 1407, 1411, 1480
Providence 340, 1376, 1399, 1627, 1882
Woonset 1432
Vermont
Bennington 263
Brattleboro 1401, 1441
Washington DC (General) 289, 290, 306, 320, 332, 355,
961, 1112, 1148, 1193, 1636, 1876
The South (General) 415, 423, 433, 438, 444, 450, 451, 461,
478, 480, 486, 488, 509, 512, 521, 529, 547, 940, 1046,
1052, 1086, 1482, 1485, 1496, 1773, 1810
Alabama (General) 1826, 1840, 1848
Birmingham 548, 1028, 1483, 1484, 1730
Florence 464
Godsden 487
Huntsville 533
Mobile 417
Montgomery 534
Tuscaloosa 421

South (cont'd)
  Arkansas (General) 1343, 1812, 1842
    Eureka Springs 523
    Little Rock 416, 446, 490, 502, 990, 1727, 1917
  Florida (General) 427, 544, 1823, 1847
    Gainesville 1918
    Miami 448, 493, 494, 543, 545, 1020, 1146, 1147
    Pensacola 489, 989
    St. Augustine 435, 436, 1115
    Tampa Bay 483, 499
    Tarpon Springs 1104
  Georgia (General) 501, 518, 520
    Atlanta 424, 432, 442, 443, 452, 457, 458, 476, 479,
      482, 485, 508, 511, 514, 526, 538, 539, 540, 541, 666,
      881, 1029, 1107, 1198, 1212, 1476, 1477, 1486, 1491,
      1494, 1724, 1732, 1743, 1744, 1745, 1912-1914, 1919,
      1920
    Augusta 431
    Dareen 473
    LaGrange 474
    Savannah 466, 471, 986, 1068
  Kentucky (General) 440a, 521a
    Lexington 1044, 1045, 1736
    Louisville 453a, 475, 490a, 527a, 650, 680, 1110, 1500,
      1723
  Louisiana (General) 1180, 1343, 1843, 1849
    New Orleans 422, 425, 437, 454, 477, 505, 517, 519, 530,
      531, 535-537, 915, 987, 1081, 1138, 1145, 1341, 1487,
      1492, 1498, 1499, 1728, 1729, 1737, 1739, 1740, 1911
    Shreveport 513, 542, 1504
  Mississippi (General) 496
    Jackson 1490
    Kociusko 418
    Vicksburg 549
  North Carolina (General) 481, 1505, 1839
    Charlotte 470, 1910
    Elizabeth City 1474
    Greensboro 428
  South Carolina
    Camden 1478
    Charleston 420, 440, 460, 465, 468, 469, 498, 507, 524,
      526a, 1161, 1479, 1497, 1734, 1735, 1738
    Florence 472
    Salem 1502
  Tennessee (General) 1831
    Clarksville 449
    Columbia 419
    Forest Park 491
    Memphis 462, 492, 515, 1741, 1742
    Nashville 445, 467, 528, 532, 1488, 1501
    Oakridge 463

Virginia (General) 453, 1025, 1062, 1373, 1932
  Alexandria 459, 510
  Appomattox 495
  Hopewell 434
  Jamestown 506
  Lynchburg 429
  Norfolk 497
  Petersburg 456, 484, 527, 1731
  Richmond 426, 430, 447, 455, 503, 504, 522, 525, 1076,
    1495, 1726
  Waterford 516
West Virginia (General) 500, 1342
  Wheeling 546
The Midwest (General) 556, 583, 601, 606, 607, 678, 691,
    1232, 1312, 1463
  Illinois (General) 563, 1313
    Canton 1314
    Chicago 550, 553, 562, 565, 570, 573, 577-580, 584, 585,
      589, 592, 597, 600, 612, 614, 619, 621, 625, 630, 633-
      635, 639-641, 649, 666, 671-676, 693, 695-697, 704,
      707, 708, 712, 718, 724, 727, 879, 883, 895, 906, 907,
      917, 923, 931, 966, 969, 981, 984, 992, 994, 1016,
      1017, 1035, 1070, 1093, 1097, 1118, 1125, 1126, 1136,
      1137, 1155-1157, 1197, 1200, 1245, 1247, 1446, 1451-
      1453, 1456, 1458, 1467, 1470, 1471, 1677, 1686, 1689,
      1696, 1699, 1700, 1702, 1703, 1708-1711, 1715, 1717,
      1718, 1722, 1892, 1895, 1898, 1905, 1906, 1908, 1909,
      1940, 1952, 1991, 2004, 2017, 2022, 2065, 2066, 2068
    Quincy 576
    Springfield 1455
  Indiana
    East Chicago 1140
    Fort Wayne 1103, 1900, 1901
    Gary 574, 656, 683, 985, 1043, 1893
    Indianapolis 557, 687, 713, 911, 1687, 1705
    Lafayette 689, 703
    Muncie 1685
    New Harmony 1443
    Terre Haute 706
  Iowa (General) 1824, 1837, 1838
    Cedar Rapids 593
    Davenport 2067
    Dubuque 1508
    Muscatine 705
  Kansas (General) 658, 731, 1825
    Kansas City 832
    Manhattan 552
    Wichita 832, 1916

Midwest (cont'd)
  Michigan (General)   653, 1004
    Ann Arbor 1442
    Battle Creek 1449
    Belding 631
    Detroit 559, 571, 596, 598, 609, 613, 620, 626, 627, 636,
      638, 661, 666, 709-711, 920, 921, 932, 952, 983, 1003,
      1018, 1042, 1053, 1061, 1077, 1162, 1167, 1209, 1315,
      1316, 1448, 1681, 1683, 1684, 1889, 1944, 1989
    Grand Rapids 667
    Grosse Pointe 632, 647
    Kalamazoo 1890, 1896
    Marine City 677
    Traverse City 1469
  Minnesota (General)   622, 690, 1083, 1317
    Duluth 1692
    Minneapolis 617, 645, 646, 958, 1318, 1319, 1690, 1691,
      1695, 2006
    St. Paul 564, 1205, 1462
    Winona 1239
  Missouri (General)   608, 1836, 1841
    Jefferson City 654
    Kansas City 575, 603, 665, 684, 685, 730, 1707
    St. Louis 595, 602, 611, 616, 637, 657, 694, 733, 735,
      1006, 1088, 1101, 1102, 1447, 1450, 1454, 1464, 1698,
      1713, 1894, 1897, 2069
    Westport 722
  Ohio (General)   567, 618, 723, 1716
    Cincinnati 581, 591, 594, 610, 615, 643, 650, 659, 660,
      668, 669, 682, 688, 699, 700, 719, 725, 728, 947, 1096,
      1217, 1461, 1472, 1701, 1706, 1714, 1720, 1891, 1903,
      1904
    Cleveland 568, 572, 670, 692, 726, 939, 1049, 1149, 1169,
      1457, 1680, 1688, 1693, 1694, 1697, 1899, 1902
    Columbus 1207, 1465, 1466
    Forest Park 681
    Marietta 1468
    Newfields 720, 1719
    Toledo 582, 1091, 1679, 1721, 1907
    Worthington 1444
  Wisconsin (General)   587, 629
    Kohler 551
    LaPointe 734
    Madison 623, 2051

Milwaukee 554, 555, 558, 560, 561, 566, 586, 588, 590, 599,
    604, 624, 648, 651, 652, 655, 663, 679, 686, 698, 701,
    702, 715, 716, 717, 729, 732, 891, 1099, 1100, 1231,
    1242, 1246, 1248, 1320-1340, 1471, 1473, 1682, 1704,
    1712, 2008
  Racine 999, 1142
The West (General) 736-740, 822, 841, 860, 863, 933, 1218
  Alaska (General) 796
    Anchorage 781
    Fairbanks 763
  Arizona
    Phoenix 849, 854, 857, 858, 875, 876, 1512, 1973
    Tucson 842, 858, 865, 873
    Yuma 856
  California (General) 742, 750, 755, 760, 765, 779, 794,
      798, 813, 814, 1850, 1943
    Alviso 757
    Auburn 2050
    Long Beach 1132
    Los Angeles 747, 752, 753, 767-772, 774, 776, 777, 784,
        789, 797, 799, 800, 803, 807, 809, 812, 816, 945, 955,
        973, 980, 987, 1078, 1113, 1122, 1131, 1135, 1143, 1195,
        1220, 1223, 1234, 1344, 1510, 1516, 1521, 1758, 1766,
        1767, 1924, 1925, 1930, 1946, 1968, 1986, 2049
    Monterey 1144
    Nevada City 2050
    Oakland 788, 1764, 1768
    Oceanside 815
    Riverside 792
    Sacramento 762
    San Diego 743, 749, 775, 785, 790, 810, 901, 975, 1960
    San Francisco 744, 754, 756, 758, 761, 764, 766, 771,
        778, 782, 783, 795, 801, 802, 805, 806, 956, 957, 963,
        965, 971, 977-979, 991, 1001, 1002, 1036, 1079, 1106,
        1150, 1151, 1154, 1165, 1170, 1181, 1184, 1188, 1506,
        1509, 1511, 1515, 1748-1752, 1754-1757, 1759, 1760,
        1769, 1926, 1928, 1929, 1937, 1980, 1990
    Santa Barbara 1131
    Santa Cruz 808
    Santa Monica 1756
    Sonoma 786
    Stockton 964, 974
    Yuba City 746
  Colorado
    Boulder 819, 836
    Denver 824, 829, 832, 833, 1202, 1240, 1489
    Fort Collins 835

West (cont'd)
  Idaho
    Paris 1493, 1513
  Montana (General) 821
    Helena 830
    Lewistown 837
  Nebraska (General) 1116
    Lincoln 1459
    Omaha 862, 871, 1445, 1460
  Nevada (General) 1345
    Las Vegas 864
    Virginia City 852, 954, 972
  New Mexico
    Albuquerque 838, 839, 858, 866, 869, 1130, 1503, 1921
    Dawson 861
    Santa Fe 846
  North Dakota (General) 825
    Jamestown 831
    New England City 820
  Oklahoma (General) 1343, 1346-1348
    Oklahoma City 1922
    Tulsa 1010
  Oregon (General) 787, 804
    Astoria 1349
    Bull Run 773
    Portland 741, 795, 811, 935, 1084, 1214, 1215, 1517-
      1520, 1761, 1762
  South Dakota (General) 1250
  Texas (General) 850, 1343, 1827, 1845
    Abilene 1133
    Beaumont 859, 1733
    Brownsville 845, 874
    Chappell Hill 1237
    Corpus Christi 870
    Dallas 840, 851
    El Paso 858, 872, 1119, 1120, 1192
    Galveston 855, 868, 1202, 1507
    Houston 843, 844, 847, 853, 867, 868, 913, 1039, 1932
    San Antonio 1098, 1117, 1121, 1481, 1725, 1746
  Utah
    Corinne 827
    Iron City 817, 1475
    La Plata 826
    Salt Lake City 823, 828
  Washington (General) 1235, 1352
    Seattle 745, 748, 751, 759, 780, 791, 1084, 1221, 1233,
      1350, 1351, 1522, 1747, 1753, 1763, 1927, 2070
    Walla Walla 834
  Wyoming (General) 818
    Cheyenne 1514
    Fort Laramie 1915

TOPICAL INDEX

This is an index to topics represented in recent research into the urban environment and which appear in the bibliography. Items listed for each topic included have been selected from the bibliography due to their broad general coverage of the topic or because they have been the focus of important specialized study.

Business Districts-Revitalization 5, 16, 32, 49, 61, 67, 74, 76, 92, 95, 97, 103, 125, 130, 151, 174, 201, 204, 218, 388, 808, 1817, 1989

Cities in Literature 64, 111, 196, 400, 554

Community Development 50, 63, 68, 92, 120, 125, 193, 204, 746

Crime and Law Enforcement 56, 57, 83, 93, 107, 112, 149, 150, 170, 207, 351, 422, 461, 564, 630, 661, 709, 710, 711, 949, 1131, 1476

Families 4, 183, 192, 248, 266, 329, 339, 383, 420, 938, 981, 1019, 1029, 1113, 1123, 1138, 1192, 1224, 1249, 1361, 1402, 1418, 1426, 1472, 1517

Federal Aid to Cities 46, 102, 105, 159, 427, 1533, 1764

Health Services 85, 89, 94, 168, 195, 228, 302, 537, 669, 679, 775, 860, 868, 893, 1190, 1237, 1248, 1692

Historic Preservation 97, 129, 133, 141, 161, 223, 573

Housing 11, 19, 23, 48, 53, 59, 68, 72, 88, 91, 92, 97, 110, 125, 136, 139, 140, 165, 171, 172, 208, 209, 217, 258, 395, 416, 423, 492, 505, 569, 610, 611, 670, 682, 779, 797, 843, 883, 984, 1147, 1320

Municipal Finances 13, 26, 38-40, 66, 86, 119, 169, 202, 205, 561, 590, 947, 1580, 1593, 1815, 1973

Municipal Services 30, 33, 71, 87, 187, 194, 213, 404, 513, 776, 832, 860, 993, 1078, 1702, 2004, 2045, 2049

Neighborhoods 21, 29, 54, 74, 148, 155, 176, 190, 210, 230, 615, 888, 1562, 1935

Planning 7, 27, 37, 44, 62, 79, 100, 122, 123, 136, 162, 163, 168, 176, 181, 182, 186, 188, 197, 199, 212, 264, 284, 361, 452, 520, 646, 678, 739, 740, 741, 745, 755, 803, 1374, 1534, 1596, 1656, 1659, 1747, 1756, 1943, 1952, 1966, 1969, 1972, 1979, 1984, 2023, 2025, 2048, 2050

Population   8, 9, 12, 14, 15, 17, 22, 28, 41, 106, 115, 131,
    138, 143, 156, 175, 180, 189, 200, 256, 381, 521, 849, 906,
    1038, 1175, 1186, 1217, 1369, 1583, 1676, 2017
Urban Economics   47, 50, 70, 78, 81, 82, 88, 118, 132, 152,
    155–157, 194, 203, 204, 219, 236, 246, 249, 341, 606, 798,
    926, 1034, 1676, 1852
Urban Geography   65, 85, 114, 164, 191, 229, 234, 558, 672,
    1563, 1636
Urban History (General)   3, 25, 42, 43, 52, 58, 59, 64, 69,
    77, 96, 113, 116, 124, 129, 134, 146, 153, 158, 160, 167,
    169, 175, 178, 179, 180, 188, 215, 216, 304, 369, 452,
    841, 1037, 1048, 1089, 1557, 1771, 1942
Urban Landscape   100, 126, 224, 227, 764, 1035
Urban Policy   1, 10, 26, 31, 63, 73, 91, 98, 99, 101, 108, 137,
    160, 214, 235, 250
Urban Sociology   18, 20, 34, 45, 51, 52, 55, 56, 65, 75, 80,
    84, 90, 109, 117, 121, 127, 135, 142, 144, 147, 148, 154,
    155, 166, 173, 177, 183, 185, 190, 206, 210, 211, 220,
    221, 225, 226, 260, 290, 316, 394, 693, 896, 1121, 1297,
    1406, 1471
Zoning   184, 470, 853, 859, 951